Lecture Notes in Energy

Volume 97

Lecture Notes in Energy (LNE) is a series that reports on new developments in the study of energy: from science and engineering to the analysis of energy policy. The series' scope includes but is not limited to, renewable and green energy, nuclear, fossil fuels and carbon capture, energy systems, energy storage and harvesting, batteries and fuel cells, power systems, energy efficiency, energy in buildings, energy policy, as well as energy-related topics in economics, management and transportation. Books published in LNE are original and timely and bridge between advanced textbooks and the forefront of research. Readers of LNE include postgraduate students and non-specialist researchers wishing to gain an accessible introduction to a field of research as well as professionals and researchers with a need for an up-to-date reference book on a well-defined topic. The series publishes single- and multi-authored volumes as well as advanced textbooks.

Indexed in Scopus and EI Compendex The Springer Energy board welcomes your book proposal. Please get in touch with the series via Anthony Doyle, Executive Editor, Springer (anthony.doyle@springer.com)

Djaffar Ould Abdeslam
Editor

Smart Meters

Artificial Intelligence to Support Proactive Management of Energy Consumption

Springer

Editor
Djaffar Ould Abdeslam
IRIMAS Laboratory
University of Haute Alsace
Mulhouse, Haut-Rhin, France

ISSN 2195-1284 ISSN 2195-1292 (electronic)
Lecture Notes in Energy
ISBN 978-3-031-27555-5 ISBN 978-3-031-27556-2 (eBook)
https://doi.org/10.1007/978-3-031-27556-2

This Springer imprint is published by the registered company Springer Nature Switzerland AG
The registered company address is: Gewerbestrasse 11, 6330 Cham, Switzerland

Preface

One of the critical factors for the transition to clean energy is the flexibility of the power grid. A flexible grid requires a constant flow of data about the network and its demand; on the other hand, clients who produce electrical power can be an active part of the demand response if they are informed about the power needs of their appliances.

"If you cannot measure it, you cannot improve it." This common management saying also holds true for the area energy efficiency. Without a clear understanding of their energy usage, consumers are unable to take steps to reduce their consumption.

In such a context, equipping buildings with smart meters is essential to improve the prediction of energy expenses within smart grids and to help end-users optimize their energy consumption. The present book describes results issued from the European Upper Rhin INTERREG project SMI (www.smi.uha.fr) which is part of a perspective linking artificial intelligence and micro-societal analysis. The book is multidisciplinary and addresses the following aspects: social, legal, environmental, and technical.

Collaboration between the partners on a cross-border scale proposes recommendations by considering the advantages and constraints of the three countries. The work is multidisciplinary and addresses the following aspects summarized in 7 chapters:

Chapter: "Advanced State of the Art Based of Smart Meters Already Carried Out in Europe and Around the World": An overview of drivers and barriers of smart meter adoption, important factors of their social acceptance as well as technology-oriented issues of smart meters on a national and international scale.

Chapter: "Understanding the Sources of Consumer Resistance to Smart Meters": Examining the underlying factors sustaining customer resistance to smart meters, this chapter shows that these devices can be the source of diverse forms of consumer disempowerment. Against this backdrop, recommendations for energy suppliers to reduce consumer disempowerment sources in the use of smart meters are developed.

Chapter: "Smart Meters Improved by NILM": An overview of the topic of artificial intelligence in the field of appliance detection will take you through the state of the art in the research area of Non-Intrusive Load Monitoring.

Chapter: "Helping Consumers to Reduce Their Energy Consumption and Greenhouse Gases Emissions: What Tool to Develop?": An in-depth analysis of consumers' representations of the energy and ecological issues to help them reducing their consumption using a web interface we are developing

Chapter: "Security Aspects of Smart Meter Infrastructures": A theoretical comparative analysis of French, German, and Swiss smart metering technologies, architecture, and protocols. Some parts of German Smart Meter Gateway penetration testing analysis.

Chapter: "Legal Aspects of the Smart Meter Rollout in Germany, France and Switzerland": An overview of the European and National regulatory framework of the smart meter rollout. It examines the general legal and data protection implementation of Directive 2009/72/EC in the Member States Germany and France and gives a brief overview of the Swiss path.

Chapter: "Technoeconomic Review of Smart Metering applications": A technoeconomic analysis of sixteen smart metering applications is carried out based on the involved stakeholders' interests and benefits, requirements, as well as challenges, barriers, and limitations. A business case is analyzed for each application, and the current status of the application is identified in Switzerland according to the feedback collected by various local industrial partners.

Mulhouse, France Djaffar Ould Abdeslam

Acknowledgements This work was supported via the European Union project "SMI Smart Meter Inclusive" (Ref: 4725/6.4), by the Program Interreg-V Upper Rhin. This work is also supported by the Swiss Confederation and the Swiss cantons of Aargau, Basel-Landschaft, and Basel-Stadt.

The editor and the authors would like to thank the Interreg secretariat, and the project manager teams at the Universities of Haute Alsace, Strasbourg, Freiburg, Furtwangen, Offenburg, Kehl, and Fachhochschule Nordwestschweiz for their support.

Contents

Advanced State of the Art Based of Smart Meters Already Carried Out in Europe and Around the World

Manuel Saroos, Ines Gavrilut, and Barbara Koch

1 Introduction

According to Wüstenhagen et al. (2007), acceptance of novel energy technologies is defined in terms of perceptions of stakeholders involved in energy projects. Sauter and Watson (2007) state that acceptance varies from passive approval with novel technologies to more active approval as for example by promoting a technology. The adoption of new technologies is defined as the action of purchasing and using a technology (Broman Toft et al. 2014), which can be measured through market share. Additionally, multiple studies include behavior towards energy technologies in their definition of acceptance. Therefore, many studies focused on different drivers, which have a strong effect on the acceptance of technologies, especially related to the subjects of technology, innovation management, and social psychology.

Within acceptance research in social psychology, the focus lies on individual user acceptance and associated individual values. Values are a fundamental part of the personality characteristics of each individual (Schwartz 1994). Moral values in an ethics of technology context are perceived characteristics of the technology (Flanagan et al. 2008). Venkatesh et al. (2012) categorized the most important factors for technology as technology-specific beliefs, social influences, and personality beliefs. They stated that technology-specific beliefs include beliefs that a technology will be useful and will improve the achievement of a consumer's objective as well as awareness towards the user-friendliness associated with a technology. In addition, consumers appear to adopt a technology more likely in case they perceive simplifying conditions, including support for the usage of a technology. Social influences such as family and friends are important drivers of perceptions. Therefore, this can lead towards using a technology in case the individuals believe that the use will enhance their social

M. Saroos (✉) · I. Gavrilut · B. Koch
University of Freiburg, Tennenbacherstr. 4, 79104 Freiburg, Germany
e-mail: manuel.saroos@felis.uni-freiburg.de

D. Ould Abdeslam (ed.), *Smart Meters*, Lecture Notes in Energy 97,
https://doi.org/10.1007/978-3-031-27556-2_1

status (Venkatesh et al. 2012). Personality-specific values and beliefs usually refer to personal norms as factors for pro-environmental behavior.

By reviewing 49 papers about acceptance due to smart grid technologies, Milchram et al. (2018) showed that moral values can be important factors for consumer acceptance of technologies such as Smart Meter. Specific moral values have a high impact on the adoption of new technologies and must be considered during the planning process of a rollout. Especially as they act as drivers or barriers of smart grid acceptance. The following chapter gives an overview of these drivers and barriers, which were examined in multiple studies. The most important factors will be reviewed in Chapter "Smart Meters Improved by NILM".

2 Barriers and Drivers of Smart Meter Adoption

In case a value provides motivation or reason to use Smart Meters or the perception of the usage of Smart Meter is to have a positive influence on this value, it is classified as a driver. If the individual's perception towards this technology is a fear of negative consequences and expressed by a specific value, it is classified as a barrier.

According to the latest studies about the acceptance of Smart Meter technologies, the drivers are environmental sustainability as well as transparency and accuracy. Privacy, security, information, (mis)trust, health, control and autonomy as well as inclusiveness were found as barriers to Smart Meter acceptance. Affordability of energy was partly identified as driver and partly as barrier (by one study). All these values arose out of studies using inductive qualitative approaches. Most of them were also included in quantitative studies, with the exception of inclusiveness and transparency.

2.1 Drivers of Smart Meter Acceptance

Customers perceived Smart Meters as a device, which helps them to save energy by improving the representation of the energy consumption so that it is possible to lower the energy costs and emissions (Guerreiro et al. 2015). Smart Meters were also seen to improve the security of supply, as they are faster in detection and reduction of power outages than conventional meters (Park et al. 2014). In the combination of Smart Meters and household electricity storage systems, it can reduce electricity supply interruptions by serving as a buffer and decoupling electricity generation from consumption. Household electricity storage systems allow to reduce the risk of supply interruptions because they can serve as a buffer for excess energy and allow to decouple electricity generation from consumption (Römer et al. 2015). Barnicoat and Danson (2015) stated that the energy saving potential of Smart Meters was seen as a driver for the technology acceptance. Guerreiro et al. (2015) found out that transparency and accuracy were additional values promoting the Smart Meter technologies, as they can help to get a better overview of energy consumption data and to contrast the consumption patterns with costs and impacts on the environment.

2.2 Barriers of Smart Meter Acceptance

Due to the introduction of Smart Meters some costumers were concerned about a loss of control and autonomy, as towards energy suppliers who could control their energy consumption (Buchanan et al. 2016; Guerreiro et al. 2015). Although the value control was mostly perceived as a barrier (Barnicoat and Danson 2015; Buchanan et al. 2016; Guerreiro et al. 2015; Krishnamurti et al. 2012), it was also stated that control enhances acceptance with regard to an automated demand-side response tariff which was clearly defined and included an option of overriding (Buchanan et al. 2016).

Health risks due to Smart Meter technology were found to be barriers (Draetta 2019; Guerreiro et al. 2015; Hess and Coley 2014; Park et al. 2014; Raimi and Carrico 2016). The exposure of electromagnetic radiation from Smart Meters was perceived to be highest health risk. Concerns about the data security was often identified to be a barrier to Smart Meter acceptance (Hess and Coley 2014; Park et al. 2014; Raimi and Carrico 2016; Zhou and Brown 2017). Connected to that were perceived threats to the consumer's privacy.

Buchanan et al. (2016) found out that increasing mistrust of the benefits of Smart Meters was linked to the consumer's fears to cover the costs of their introduction to the market whereas only the energy providers would financially profit. Similar to that, Hall et al. (2016) stated that perceptions of the consumers were to be responsible for saving energy while lowering the consumer prices wouldn't be implemented by energy suppliers. Lastly, consumers were also concerned about missing inclusiveness and feared that people with disabilities, lower affinity to IT systems or older people would be neglected in the run-up to the introduction of Smart Meters and smart grid technologies (Buchanan et al. 2016).

The examined social values, how they were perceived concerning Smart Meters or Smart Meter technologies and which studies analyzed those effects are summarized in Table 1.

3 Important Factors of Social Acceptance Towards Smart Meter

3.1 Rollouts and Information

Sareen (2020) compared national Smart Meter rollouts in Norway (the Smart Meter rollout was completed in Norway, with over 97% coverage by January 2019) and

Table 1 Effects of Smart Meters on examined social values

Social values	Smart Meter effect	Sources
Environmental sustainability	**Driver**	Guerreiro et al. (2015), Zhou and Brown (2017)
Transparency and accuracy	**Driver**	Guerreiro et al. (2015)
Affordability of energy	**Driver**/*barrier*	Barnicoat and Danson (2015)
Privacy	*Barrier*	Bolderdijk et al. (2013), Buchanan et al. (2016), Chen et al. (2017), Chou and Yutami (2014), Fredersdorf et al. (2015), Guerreiro et al. (2015), Hess and Coley (2014), King and Jessen (2014), Krishnamurti et al. (2012), Mah et al. (2012), Wunderlich et al. (2012), Zhou and Brown (2017)
Information	*Barrier*	Hellmuth and Jakobs (2020), Sareen (2020), Sareen and Rommetveit (2019)
Security	*Barrier*	Hess and Coley (2014), Park et al. (2014), Raimi and Carrico (2016), Zhou and Brown (2017)
Trust	*Barrier*	Buchanan et al. (2016), Chen et al. (2017), Gerpott and Paukert (2013), Hall et al. (2016)
Health	*Barrier*	Draetta (2019), Guerreiro et al. (2015), Hess and Coley (2014), Park et al. (2014), Raimi and Carrico (2016)
Control and autonomy	*Barrier*	Barnicoat and Danson (2015), Buchanan et al. (2016), Guerreiro et al. (2015), Krishnamurti et al. (2012)
Inclusiveness	*Barrier*	Buchanan et al. (2016)

Portugal due to their effects on social acceptance. Both countries were predominant positive about a Smart Meter rollout. Although there were several different results for these countries, the similarities were that the control for social and technical aspects rest mainly at the national scale, whereas the responsibility for social aspects is related to the household scale. This is different to the technical aspects, where the responsibility remains on a national scale. In general, controlling how Smart Meters were introduced, what characteristics should be delegated, and how they should interact with energy use, has remained constant on a national scale in both countries. Norwegian participants felt excluded from any real influence on this (Sareen and Rommetveit 2019), while in Portugal, control remained in the hands of the experts and therefore has not been considered as a part of a wider social concern.

In Portugal, the social aspects were fundamentally missing from the debate, which remained confined to technical matters and largely outside the public eye, since rolling out the program was not mandatory and DSO absorbed the costs. Respondents in Portugal saw the Smart Meter rollout as a deeply contingent process. Energy researchers pointed out that the public perception that Smart Meters would increase costs made rollout difficult for the DSO, which had to invest in deploying this infrastructure.

Norwegian study participants reported several concerns. Some did not feel they could trust energy companies with data on their electricity use. Others argued that Norway's investments in energy resilience were only helping other countries (for example to reduce Germany's dependence on coal) with no greater advantages for Norway. The eagerness and ability of participants to reduce their electricity consumption varied based on the Smart Meter simulation, but they shared a general sense of deprivation. They were unable to influence how Smart Meters were introduced, what this means for their energy future, or how their data was used or misused. Sareen (2020) concluded that embedding social and technical differentiation due to such scalar biases risks dehumanising technical aspects while detechnicising social aspects in this early intersection of energy transitions and automation.

Another Study concerned about rollout strategies in Germany due to the awareness of Smart Meters in the public (Hellmuth and Jakobs 2020). Hellmuth and Jakobs (2020) stated that a good deployment needs a high degree of acceptance among end-users. In particular with regard to the technology's data privacy and data security aspects. To research that question, they explored in interviews the attitudes of people towards Smart Metering. Thereby, they focused on the needs for information, connectivity and information, data protection and data security. Their results showed that the highest level of demand for information is about advantages and disadvantages of these systems (see Fig. 1). Also, general functions, tariff models and technical aspects of the systems needed more clarification.

Informedness is defined as the extent of perceived availability of knowledge on certain topics or subjects. Therefore, it is distinguished from factual knowledge (Hellmuth and Jakobs 2020). Renn (2015) stated that knowledge and information alone does not directly lead to acceptance. As acceptance studies highlight, attitudes toward technologies are often influenced by self-assessments and emotions or feelings and can therefore be an important component of acceptance (Park et al. 2014; Renn 2015).

The need to inform end consumers sufficiently about Smart Metering systems are underlined by various authorities (Verbraucherzentrale Bundesverband (VZBV)

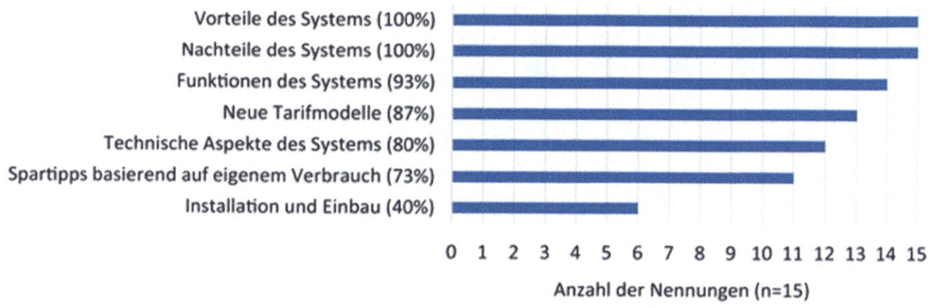

Fig. 1 Need for information for Smart Meter background (Hellmuth and Jakobs 2020)

2016). Two directions of informing were differentiated, the active "information-pull" ("I want to inform myself") and the passive "information-push" ("I want to be informed") (Mauelshagen and Jakobs 2016). The results for "information-pull" (see Fig. 2) showed that web sites are considered to be the most important formats, whereas for "information-push" (see Fig. 3) flyer and information letters are preferred.

Hellmuth and Jakobs (2020) pointed out that there are strong information deficits and information needs regarding measurement systems and rollout. They also stated that data privacy and data preservation are particularly important and are objectively regarded. Insecurity and potential threats concern the use of data in particular. Cautious communication and accompaniment that include the preferences

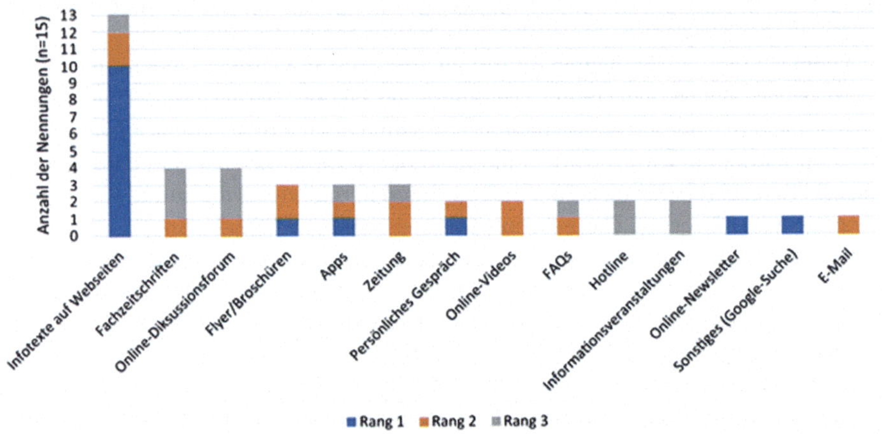

Fig. 2 Most important formats for "information-pull" (Hellmuth and Jakobs 2020)

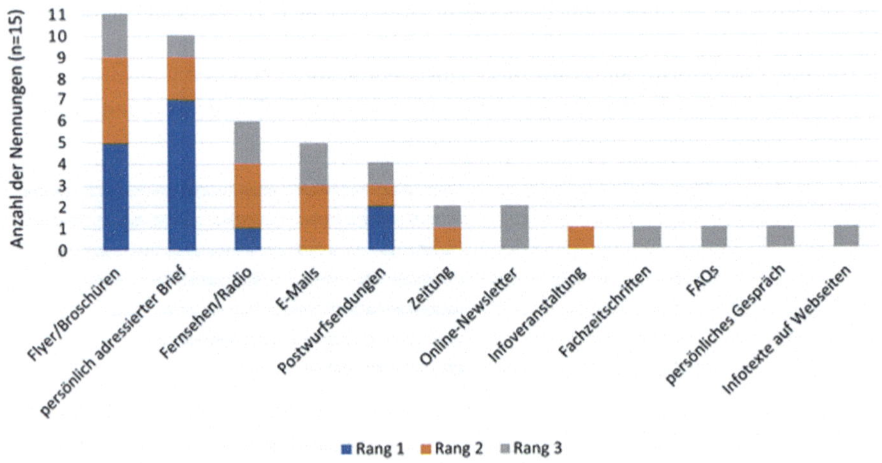

Fig. 3 Most important formats for "information-push" (Hellmuth and Jakobs 2020)

and the living environment of the public concerns is important for Smart Meter implementation.

Several studies showed that there is still a lack of understanding of what Smart Meters are. A study of Initiative D21 e. V. (2018) showed that only 10% of the participants were familiar with the term Smart Meter. In another study from Hitschfeld. Büro für strategische Planung (2017), the percentage was enhanced so that about 34% were familiar with it.

3.2 *Health*

Some studies focused on social acceptance of Smart Meters due to possible health issues. Draetta (2019) focused on the public controversy over Smart Meters, especially the Linky electricity meter in France by analyzing the published French Press and several institutional documents. This started shortly after the national rollout in France at the end of 2015. One main topic, which evolved first from the public controversy, was the human exposure to the electromagnetic field (EMF) emitted by the meter and associated risks for the human health and physical comfort. The time scale of the media coverage in Fig. 4 shows how the specific areas of concern emerged over time.

Anses (Agence nationale de sécurité sanitaire de l'alimentation, de l'environnement et du travail) (2016) concluded that EMF emissions of Linky and other Smart Meters result in low levels of exposure compared to the regulatory limits. They also stated that short- or long-term health effects due to exposure to the EMF emissions of the meters are very unlikely. Based on the low levels of exposure, short-term danger could not be detected. Due to the absence of specific scientific literature about health effects linked to the PLC frequency range, there is no possible

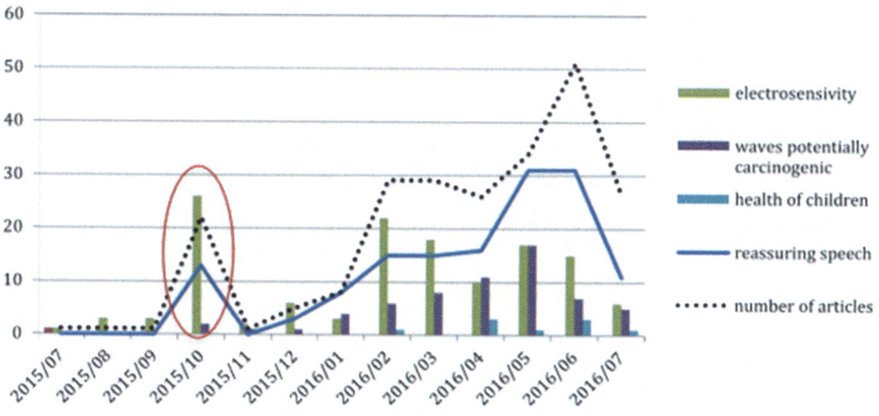

Fig. 4 Media coverage of possible health issues due to Smart Meters in France (Draetta 2019)

conclusion about the long-term danger. Although the intension of these results were to reassure the public, it also generated more doubts due to the scientific uncertainty about Smart Meters.

Although the media coverage was low in quantity, the health issue played a key role in opening up the controversy to new categories of opponents (Fig. 5). As this public controversy started as a health controversy due to human exposure to the EMF emissions of Smart Meters, it included a broad range of issues as invasion of privacy, security, cost and environmental impact. It therefore shows the social dimensions of Smart Meters as a technical infrastructure which generates for each technical aspect (remote communication (PLC), collection of consumption data, automation of tasks etc.) relating social issues. Draetta (2019) concluded that the public opposition was driven by missing trust in the rollout as it was seen to be governed by economic and corporate concerns rather than by a common environmental policy.

In addition, Mengolini and Vasiljevska (2013) already stated in their report for the European Commission about the social dimensions of Smart Grids, that despite evidence to date suggests that exposure to electromagnetic radiation produced by Smart Meter devices do not pose any risk to health, the fears of the consumers about health impacts have to be seriously considered. Fears about possible health effects could lead to rollout delays or even threaten the installation process (Neuburg 2013).

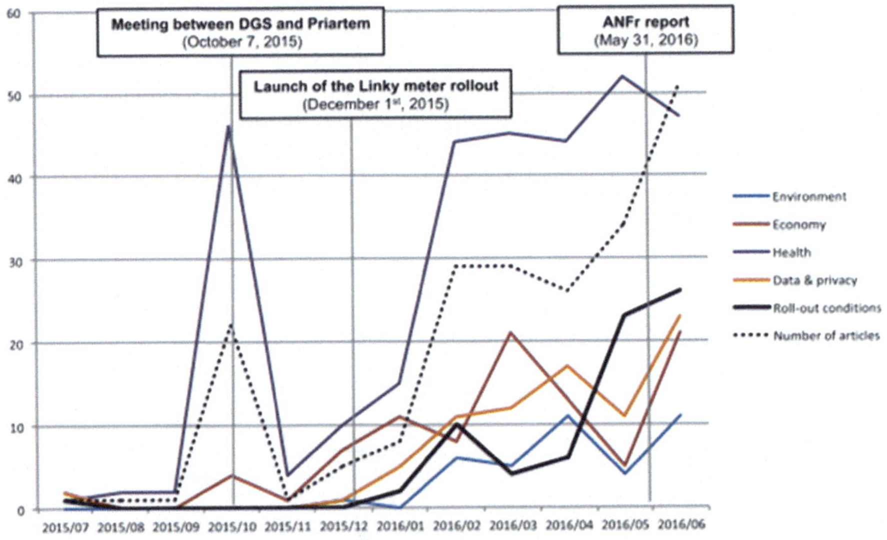

Fig. 5 Media coverage of the public controversy in France according to the main key issues (Draetta 2019)

4 Technology-Based Factors of Smart Meter

Zhou and Brown (2017) summarized the functionalities of a Smart Meter as the capabilities to meter and to communicate. This leads to the main goals of Smart Grids (SG) by using Advanced Metering Infrastructure (AMI) to measure, communicate and analyze the consumption data of the users. The both-sided communication of Smart Meters and users as well as Smart Meters and energy providers, allows informing users about their consumption habits, improves maintenance, as well as demand and expansion management. The consequential large amount of data traffic needs efficient data management (see also Fig. 6).

Guerreiro et al. (2015) considered the two-way communication as the integral part of Smart Meters and the main advantage over regular meters. Besides the already mentioned communication ways, it also implies communication between users. Therefore Fadel et al. (2015) stated that the independence of the communication module is crucial as it allows Smart Meters to measure and store data, even though the connection would be interrupted.

There are two options to establish communication: wired and/or wireless connections. Whereas wired technologies provide higher bandwidth and reach longer distances, wireless networks could be less expensive and may reach difficult areas (Avancini et al. 2019). A newer approach of communication with high expectations is the Internet of Energy (IEN) where energy networks transfer data through the internet. The rapid development of the Internet, as for example their data transfer capacity, will directly improve the functionality of Smart Meters (Fadel et al. 2015).

Smart Meters must be capable of sending data and additionally of receiving orders. This requires an Advanced Metering Infrastructure (AMI) with a corresponding communication network. The huge number of clients and meters producc to the need

Fig. 6 Illustration of Smart Meter systems (Avancini et al. 2019)

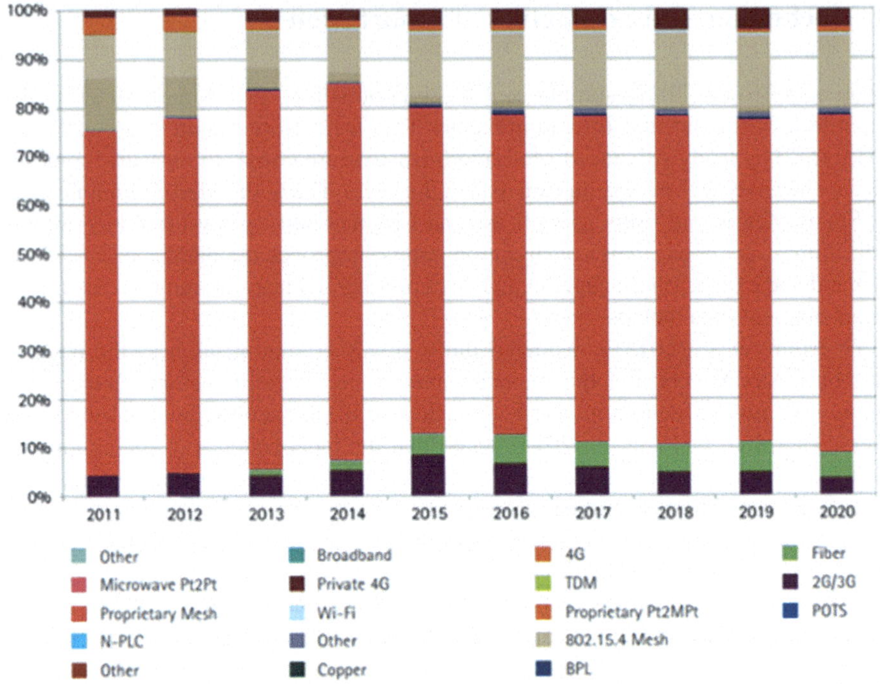

Fig. 7 Communication technologies development (Uribe-Pérez et al. 2016)

of a solid correspondence framework, fit for moving significant volumes of information. The requirements for such a network amongst others are data protection, high capacity, processing of consumption data, profitability, and development capacity. Wireless and wired technologies are both considered for those purposes (Avancini et al. 2019). Finally, it is believed that a combination of different communication technologies is currently the best way to fulfill all the demands. Therefore, the following summarizes the present situation of the available communication technologies and networks (see also Fig. 7).

4.1 Communication Technology

Bluetooth

Bluetooth, which is most present on smartphones or other portable devices, is a wireless low-cost technology with a short range of the signal. It is mainly used in Home Area Networks (HANs), with a frequency at 2.45 GHz and a bandwidth of up to 3 Mbps. Cecilia and Sudarsanan (2016) stated that this could be used for a wireless and local access in Smart Meters or other SG components.

ZigBee

ZigBee is another low-cost and wireless technology. Its frequency ranges up to 100 m. It is used in communicative home appliances and can also be applied for Smart Meter communication. It is offered for three different frequencies (868 MHz, 915 MHz, or 2.4 GHz) and allows data exchange up to 250 kbps. For some system variations as ZigBee Smart Energy Profile (SEP) wireless mesh networks can be developed through using each network node as a wireless router. Therefore, the range can be extended through the number of the nodes in the mesh (Cecilia and Sudarsanan 2016; Yi et al. 2011).

Wi-Fi

For short distance communication with a maximum range of 250 m, Wi-Fi could be also used. It allows data transfer with rates up to 600 Mbps. The uses the 2.4 or 5 GHz frequency bands. Its disadvantage is that is doesn't fulfill the requirements of low-energy consumption (Cecilia and Sudarsanan 2016).

WNAN/Wireless Ethernet (IEEE 802.11)

WNAN or wireless Ethernet is a family of standards whose technology has a high reliability and availability. Though, the data rate is affected by electromagnetic interferences which also leads to possible disturbances of other nearby devices (IEEE 802.11™ Wireless LANs 2020).

Radio Frequency (RF)

RF technology exchanges measurements and other data by wireless radio from the Smart Meters to a collection point. RF mesh is most popular topology where there is a communication between the Smart Meters and thereby form a Local Access Network (LAN) cloud to a collector (Edison Electric Institute 2011). It advantages are a large bandwidth and a self-healing characteristic of the network as the communication signals try alternative routes through active nodes in case nodes are dropping out (Gungor et al. 2011).

GPRS

The open standard technology GPRS is very effective and reliable. Tests in Ireland of Smart Meters using GPRS showed good results with success rates of 97.89% for first-time reads as well as easy deployment (Commission for Energy Regulation 2011). Disadvantageous is the moderate data rate.

Cellular Networks

Cellular networks provide another possibility. They offer a large coverage and significant data rates. Gungor et al. (2011) stated that it could be utilized to connect Smart Meters, nodes or other smart technology elements. Existing cellular networks could be used to create WANs for Smart Grids (Cecilia and Sudarsanan 2016).

PLC (Powerline Communication)

It is possible to use the existing power lines as a wired technology to transmit data with a rate up to 3 Mbps. There are challenges due to the signal noise of power lines but they have the advantage of an already existing widespread network and therefore of lower installation costs. SGs already PLC to create HANs (Gungor et al. 2011).

DSL/Optical Fiber

DSL/Optical Fiber provide high-speed data exchange by using the telephone network or optical fibers. They can be used for connecting SG elements in HANs and WANs. They also can be used to interconnect Smart Control centers. The advantages of this technology are low costs and the high bandwidth (Cecilia and Sudarsanan 2016).

Euridis

Euridis technology is the only existing standardized interface for Smart Metering technologies, which uses twisted pair cable as wired data transfer. Several Smart Meters in France are provided with this interface. Euridis is cost-effective technology, which offers the concurrent access to over 100 Smart Meters, which are connected in the same bus (Electa 2010; IEC 62056-31 1999).

4.2 Communication Networks

Smart communication networks can be distinguished between Home Area Networks (HANs), Neighborhood Area Networks (NANs) and Wide Area Networks (WANs) (see Fig. 8).

HANs usually connect several home devices such as Smart Meters, electric vehicles and others. It collects consumption pattern data as well as data about energy usage and relies on low-energy communication technologies such as ZigBee and Bluetooth or uses Wi-Fi (Fadel et al. 2015).

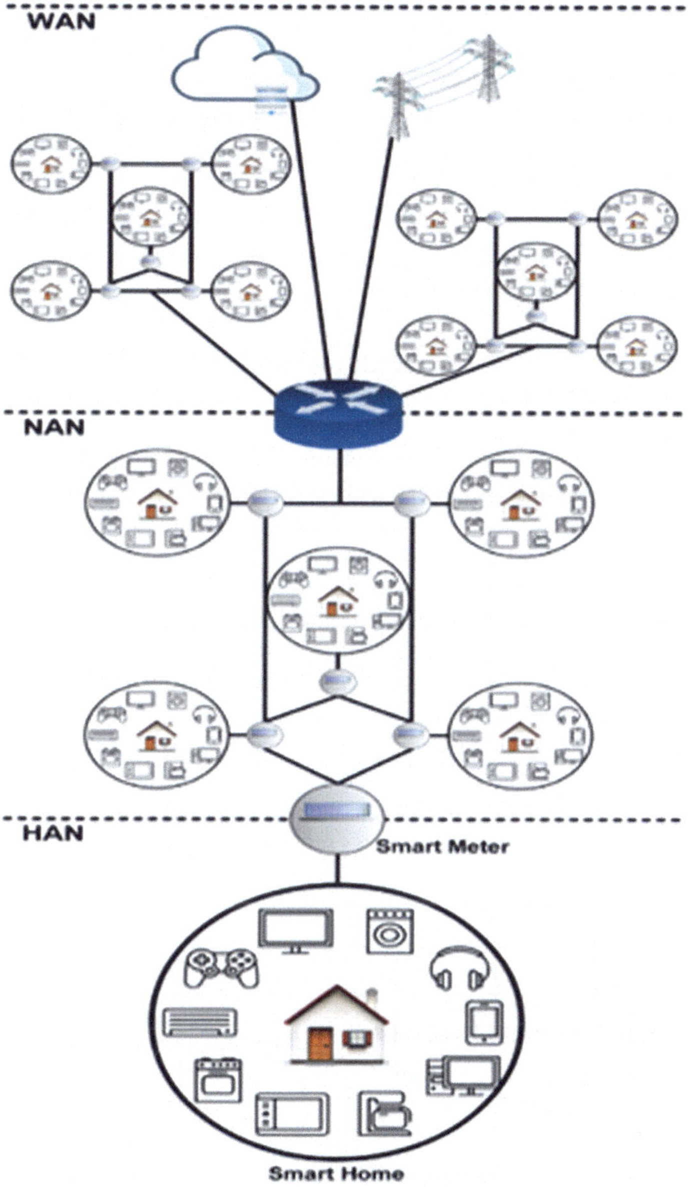

Fig. 8 Illustration of a SG network (Avancini et al. 2019)

NANs are using data collectors to link data from multiple households and Smart Meters. This requires communication technologies with a short range as Wi-Fi or RF Mesh modules (Fadel et al. 2015).

WANs connect NANs and other devices with data centers. To cover large areas wide-range and high-capacity communication techniques as cellular networks, optical fiber and PLC can be used (Fadel et al. 2015; Yigit et al. 2014).

4.3 Smart Metering Worldwide

Worldwide, around one billion Smart Meters are estimated to be installed by 2022 (Strother and Lockhart 2014). The evolution of Smart Meters deployment till 2023 is illustrated in Fig. 9.

The Council of European Energy Regulators stated that Europe has a challenging situation due to a missing common standard for Smart Meters as well as the lack of interoperability (Council for European Energy Regulators (CEER) 2013). This fact is believed to be disadvantageous concerning customer services and to prevent the market penetration of SM suppliers due to the inflictions of local and national standards, which would require certain modifications of the product (Alejandro et al. 2014). The European Commission stated that the most challenging functionality refers to consumption data frequency and the providing this data to consumers as well as third parties (COM 2014).

This leads to questions about data protection and security. It highlights the needs for a specific framework and customer data protection to be the major goals towards the development of standards. In addition, the communication technology is a major challenge for the future, because nowadays most of the Smart Metering networks use

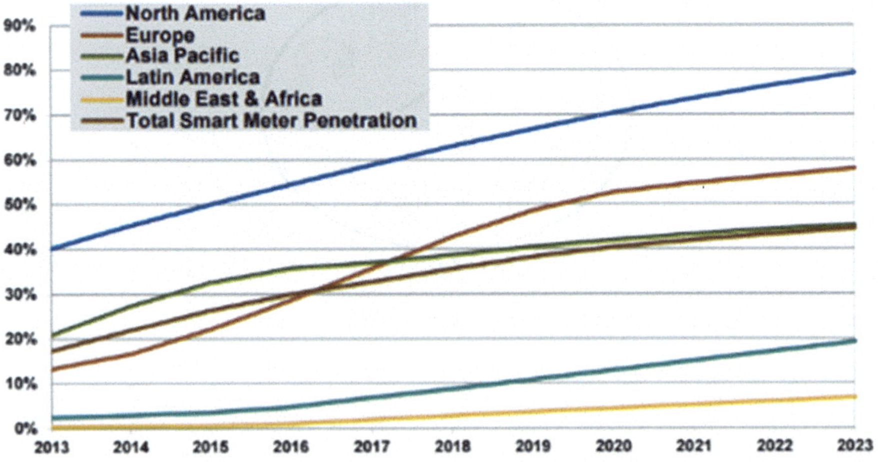

Fig. 9 Estimations of the Smart Meters deployment (Strother and Lockhart 2014)

low to medium bandwidth generating high data traffic and limiting the data transmission rate (Alexander 2007). Technical innovations to reduce data retransmissions plus the establishment of hierarchy levels with priorities due to the type of data, such as measure, control, management and others are mentioned to be helpful to increase efficiency of the communication technology in Smart Meters. Another topic is the problem with interferences. Other near-by emitting devices can affect wireless technologies, whereas noise from the power cable and other connected devices interfere with the signal in PLC. These problems can be solved by establishing new robust protection codes, transforming emitting devices and by creating legal guidelines concerning interferences (Bartak and Abart 2013).

Finally, to speed up the deployment of Smart Metering systems in Europe, the European Commission has listed several propositions for the next steps as follows (Uribe-Pérez et al. 2016):

- **Gain the trust and confidence of consumers**: Customers must be convinced of three main aspects: their rights as users, the advantages of implementing SMs, and their inclusion in demand response systems.
- **Achieve an innovative energy services market**: Synergies with the ICT industry would be critical in supporting an innovative energy economy.
- **Sensitive data protection**: The European Commission and member states must determine if new data privacy and protection system legislation is needed.
- **Data management**: Utilities and the ICT industry will have to collaborate and explore data storage possibilities.
- **Smart Meter functions**: Member states will be able to recognize similar ways of achieving cost efficiencies and ensuring fit-for-purpose in their Smart Metering rollouts according to technical and commercial interoperability.
- **Long-term economic assessment of costs and benefits**: An analysis of the essential parameters used in national rollouts, as well as the decisions made, would aid in the refinement of technology choices.

References

Alejandro L, Blair C, Bloodgood L, Khan M, Lawless M, Meehan D, Schneider P, Tsuji K (2014) Global market for smart electricity meters: government policies driving strong growth. Working paper. US International Trade Commission

Alexander B (2007) Smart meters, demand response and "real time" pricing: too many questions and not many answers

Anses (Agence nationale de sécurité sanitaire de l'alimentation, de l'environnement et du travail) (2016) Exposition de la population aux champs électromagnétiques émis par les 'compteurs communicants'. Avisrevisé de l'avis de décembre 2016, 7 Juin 2017. Anses, Maisons-Alfort

Avancini DB, Rodrigues JJPC, Martins SGB, Rabêlo RAL, Al-Muhtadi J, Solic P (2019) Energy meters evolution in smart grids: a review. J Clean Prod 217(10):702–715. https://doi.org/10.1016/j.jclepro.2019.01.229

Barnicoat G, Danson M (2015) The ageing population and smart metering: a field study of householders' attitudes and behaviours towards energy use in Scotland. Energy Res Soc Sci 9:107–115

Bartak GF, Abart A (2013) EMI of emissions in the frequency range 2 kHz–150 kHz. In: Proceedings of the 22nd international conference on electricity distribution (CIRED), Stockholm, Sweden, 10–13 June 2013

Bolderdijk JW, Steg L, Postmes T (2013) Fostering support for work floor energy conservation policies: accounting for privacy concerns. J Organ Behav 34(2):195–210

Broman Toft M, Schuitema G, Thøgersen J (2014) Responsible technology acceptance: model development and application to consumer acceptance of smart grid technology. Appl Energy 134:392–400

Buchanan K, Banks N, Preston I, Russo R (2016) The British public's perception of the UK smart metering initiative: threats and opportunities. Energy Policy 91:87–97

Cecilia AA, Sudarsanan K (2016) A survey on smart grid. In: International conference on emerging trends in engineering, technology and science (ICETETS). IEEE, pp 1–7

Chen C, Xu X, Arpan L (2017) Between the technology acceptance model and sustainable energy technology acceptance model: investigating smart meter acceptance in the United States. Energy Res Soc Sci 25:93–104

Chou J-S, Yutami GAN (2014) Smart meter adoption and deployment strategy for residential buildings in Indonesia. Appl Energy 128:336–349

COM (2014) Cost-benefit analyses & state of play of smart metering deployment in the EU-27. Commission staff working document, 356 p

Commission for Energy Regulation (2011) Electricity smart metering technology trials findings report

Council for European Energy Regulators (CEER) (2013) Status review of regulatory aspects of smart metering

Draetta L (2019) The social construction of a health controversy. The case of electricity smart meters in France. Ann Telecommun 74(1–2):5–15

Edison Electric Institute (2011) Smart meters and smart meter systems: a metering industry perspective. URL: http://www.eei.org/issuesandpolicy/grid-enhancements/documents/smartmeters.pdf. Accessed 6 Jan 2016

Electa (2010) Study on smart meters from the angles of the consumer protection and the public service obligations

Fadel E, Gungor VC, Nassef L, Akkari N, Malik MA, Almasri S, Akyildiz IF (2015) A survey on wireless sensor networks for smart grid. Comput Commun 71:22–33

Flanagan M, Howe DC, Nissenbaum H (2008) Embodying values in technology: theory and practice. In: Van Den Hoven J, Weckert J (eds) Information technology and moral philosophy. Cambridge University Press, New York, NY, pp 322–353

Fredersdorf F, Schwarzer J, Engel D (2015) Die Sicht der Endanwender im Smart Meter Datenschutz. DuD (Datenschutz Datensich) 39(10):682–686

Gerpott TJ, Paukert M (2013) Determinants of willingness to pay for smart meters: an empirical analysis of household customers in Germany. Energy Policy 61:483–495

Guerreiro S, Batel S, Lima ML, Moreira S (2015) Making energy visible: sociopsychological aspects associated with the use of smart meters. Energy Effic 8:1149–1167

Gungor V, Sahin D, Kocak T, Ergüt S, Buccella C, Cecati C, Hancke G (2011) Smart grid technologies: communications technologies and standards

Hall NL, Jeanneret TD, Rai A (2016) Cost-reflective electricity pricing: consumer preferences and perceptions. Energy Policy 95:62–72

Hellmuth N, Jakobs EM (2020) Informiertheit und Datenschutz beim Smart Metering. Z Energiewirtsch 44(1):15–29

Hess DJ, Coley JS (2014) Wireless smart meters and public acceptance: the environment, limited choices, and precautionary politics. Public Underst Sci 23:688–702

Hitschfeld. Büro für strategische Planung (2017) Akzeptanz von Technik und Technologie; Welle 1/2017: "smart meter—smart metering"
IEC 62056-31:1999 withdrawn. https://webstore.iec.ch/publication/
IEEE 802.11™ Wireless LANs. URL: https://standards.ieee.org/standard/802_11-2020.html. Accessed 22 Aug 2020
Initiative D21 e. V. (2018) Digital index 2017–2018. Jährliches Lagebild zur digitalen Gesellschaft
King NJ, Jessen PW (2014) Smart metering systems and data sharing: why getting a smart meter should also mean getting strong information privacy controls to manage data sharing. Int J Law Inf Technol 22:215–253
Krishnamurti T, Schwartz D, Davis A, Fischhoff B, de Bruin WB, Lave L, Wang J (2012) Preparing for smart grid technologies: a behavioral decision research approach to understanding consumer expectations about smart meters. Energy Policy 41:790–797
Mah DN, van der Vleuten JM, Hills P, Tao J (2012) Consumer perceptions of smart grid development: results of a Hong Kong survey and policy implications. Energy Policy 49(2012):204–216
Mauelshagen C, Jakobs EM (2016) Science meets public—customized technology research communication. In: Proceedings of the IEEE ProComm 2016: communicating entrepreneurship and innovation, Texas
Mengolini A, Vasiljevska J (2013) The social dimension of smart grids. Consumer, community, society. Publications Office (EUR, Scientific and Technical Research Series, 26161), Luxembourg
Milchram C, van de Kaa G, Doorn N, Künneke R (2018) Moral values as factors for social acceptance of smart grid technologies. Sustainability 10(8):2703
Neuburg S (2013) Smart grids: future-proofed for consumers? Consumer Futures, June 2013
Park CK, Kim H-J, Kim Y-S (2014) A study of factors enhancing smart grid consumer engagement. Energy Policy 72:211–218
Raimi KT, Carrico AR (2016) Understanding and beliefs about smart energy technology. Energy Res Soc Sci 12:68–74
Renn O (2015) Akzeptanz und Energiewende. Bürgerbeteiligung als Voraussetzung für gelingende Transformationsprozesse. Jahrb Christl Sozialwiss 56:133–154
Römer B, Reichhart P, Picot A (2015) Smart energy for Robinson Crusoe: an empirical analysis of the adoption of IS-enhanced electricity storage systems. Electron Mark 25:47–60
Sareen S (2020) Social and technical differentiation in smart meter rollout: embedded scalar biases in automating Norwegian and Portuguese energy infrastructure. Humanit Soc Sci Commun 7(1):1025
Sareen S, Rommetveit K (2019) Smart gridlock? Challenging hegemonic framings of mitigation solutions and scalability. Environ Res Lett 14:075004
Sauter R, Watson J (2007) Strategies for the deployment of micro-generation: implications for social acceptance. Energy Policy 35:2770–2779
Schwartz SH (1994) Are there universal aspects in the structure and contents of human values? J Soc Issues 50:19–45
Strother N, Lockhart B (2014) Smart meters. smart electric meters, advanced metering infrastructure, and meter communications: global market analysis and forecasts. Navigant Research, Boulder, CO
Uribe-Pérez N, Hernández L, de La Vega D, Angulo I (2016) State of the art and trends review of smart metering in electricity grids. Appl Sci 6(3):68. https://doi.org/10.3390/app6030068
Venkatesh V, Thong J, Xu X (2012) Consumer acceptance and use of information technology: extending the unified theory of acceptance and use of technology. MIS Q 36:157–178
Verbraucherzentrale Bundesverband (VZBV) (2016) Neue Stromzähler werden ab 2017 schrittweise Pflicht. https://www.vzbv.de/pressemitteilung/neue-stromzaehler-werden-ab-2017-schrittweise-pflicht. Zugegriffen: 18 Dez 2020
Wunderlich P, Veit D, Sarker S (2012) Examination of the determinants of smart meter adoption: an user perspective

Wüstenhagen R, Wolsink M, Bürer MJ (2007) Social acceptance of renewable energy innovation: an introduction to the concept. Energy Policy 35:2683–2691

Yi P, Iwayemi A, Zhou C (2011) Developing ZigBee deployment guideline under WiFi interference for smart grid applications. IEEE Trans Smart Grid 2(1):110–120

Yigit M, Gungor VC, Tuna G, Rangoussi M, Fadel E (2014) Power line communication technologies for smart grid applications: a review of advances and challenges. Comput Netw 70:366–383

Zhou S, Brown MA (2017) Smart meter deployment in Europe: a comparative case study on the impacts of national policy schemes. J Clean Prod 144:22–32

Understanding the Sources of Consumer Resistance to Smart Meters

Virginie Schweitzer and Françoise Simon

1 Introduction

Today, the combined effect of the development of renewable energy and the electrification of the whole economy, including vehicles, is radically transforming the operation of the grid. As a result, demand-side management is increasingly attracting policy attention as a tool to increase the flexibility of electricity systems and reduce the carbon intensity of electricity supply (Parish et al. 2020). Specifically, demand-side management, which is active management by the user of their own demand, is expected to result in two main consumption practices (Batalla-Bejerano et al. 2020): (1) load shifting—shifting consumption from peak to off-peak periods and, (2) strategic conservation—performing activities that reduce energy consumption. In this context, smart meters, as they are implemented in households at a large scale, constitute the cornerstone of demand-side management while improving the efficiency and reliability of the overall electricity system.

Large smart metering roll-out programs are under way or being planned throughout the developed countries and especially, in Europe. It is expected that by 2024 nearly 77% of European consumers will have a smart meter for electricity. By replacing traditional electromechanical meters, smart meters, or the mobile applications connected to them, provide customers with real-time consumption data that can be enriched with social benchmarks or energy reduction tips. Based on these forms of intelligent feedback, consumers are expected to manage their demand more effectively and change their electricity consumption patterns throughout the day (Frederiks et al. 2015). However, recent surveys in European countries highlight disappointing levels of consumer behavioral engagement with the device while there is no significant reduction in household energy consumption. As a result, a growing

V. Schweitzer (✉) · F. Simon
University of Haute Alsace, Colmar, France
e-mail: virginie.schweitzer@uha.fr

D. Ould Abdeslam (ed.), *Smart Meters*, Lecture Notes in Energy 97,
https://doi.org/10.1007/978-3-031-27556-2_2

body of literature questioning the optimism related to smart meter feedback has emerged.

Therefore, the gap between expected and actual acceptance of smart meters calls for investigations of the underlying factors sustaining customer resistance to smart meters. By examining the consumers' concerns associated with smart meters, the present chapter shows that smart meters can be the source of diverse forms of consumer disempowerment. Against this background, recommendations for a better acceptance are formulated.

2 The Ambivalence and Limited Engagement of European Consumers Towards Smart Meters

Most conceptual models that aim to explain the adoption of smart meters are based on the assumption of a rational consumer. According to this assumption, consumption feedback will increase awareness and knowledge of energy consumption levels and patterns, which will encourage households to make informed decisions to reduce their consumption, for economic and/or environmental reasons. In particular, corporate entities' strategies that provide information about the environmental impact of individual activities are seen as effective to encourage conservation behavior.

This line of reasoning was supported by some previous empirical research, where an early review of the effectiveness of feedback strategies by Darby (2006) suggested average savings in electricity consumption of around 5–15%. However, as pointed out by Nilsson et al. (2018), the optimism of feedback to reduce energy consumption has diminished. For example, a systematic review by Delmas et al. (2013) demonstrated that the more optimistic quantitative results reported in previous case studies on energy feedback came from less robust studies that suffered from methodological flaws. They observe that savings were down to 2% for the studies of the highest quality that include a control group as well as weather and demographics controls. Other studies looking at the effect of smart meter feedback have been criticized for not being able to differentiate between the effects of feedback, self-selection bias of the participants and/or the Hawthorne effect (participants behaving differently because they know they are taking part in a study) (Buchanan et al. 2016). Consistently, European Commission (2020) reported in 2017 that across the EU, smart meters result on average in 3% energy savings, which is far below the levels previously expected.

While empirical evidence indicates that the roll-out of smart meters to a large scale in Europe fails to support substantial moves in energy conservation, recent research indicates that smart meters also face a growing skepticism not only from final users, but also from intermediary actors. To illustrate, research examining British public's acceptance of smart meters points out ambivalent attitudes or apathy rather than overwhelming support, with a majority of them indicating that they are undecided about whether they should be installed in every UK home (Buchanan et al. 2016). On the other hand, targeting households' segment of particular relevance, that is, high-income and highly educated households, which are considered as early adopters of smart grid technologies, a field trial in Sweden shows that even if smart energy feedback may lead to increased awareness of energy consumption, as well as increased home comfort, several obstacles for energy consumption behavioral change can be identified (Nilsson et al. 2018). Additionally, the roll-out of smart meters may sustain active resistance among social actors. Resistance to smart meters takes various forms (Mela et al. 2018), both individual (people preventing installers from gaining access to their traditional meter to replace it) and collective (ad hoc groups or associations highlighting the pointlessness of the replacement). More surprisingly, in France, municipalities emerge as an important level of resistance against the smart meter (Chamaret et al. 2020). Urged on by local citizens, who felt they were too small to fight the introduction of smart meters, municipalities took official decisions about the roll-out on their territories, that range from letting people decide for themselves whether they want a new meter, to a total ban on installation of smart meters.

Overall, these findings reveal the ambivalence of consumers' attitudes towards the roll-out of smart meters and their limited engagement towards the device. They call into question the relevance of adoption models that postulate a proximal link between the improved knowledge of one's own energy consumption enabled by the smart meter and energy reduction behavior. Moreover, they highlight the need to understand the structure of social representations relating to smart meters, particularly those that reflect a loss of consumer power.

3 Sources of Consumer Disempowerment in the Social Representations of Smart Meters

In marketing research, the concept of consumer psychological empowerment has been enlisted as a theoretical means to address consumer quest for power through the use of an artefact (Schweitzer and Simon 2021). Specifically, psychological empowerment is intrinsically related to the notion of power as it describes an individual's internal state of enabling himself or herself to act on his or her own to reach his or her self-defined goals (Menon 2001; Spreitzer 1995). The construct integrates a cognitive core that encompasses the beliefs that individuals have regarding their competence, autonomy, and the outcomes that their activities may have on themselves and their community (Menon 2001). Conversely, consumer internal beliefs of a loss of power can be referred to as psychological disempowerment.

To collect data for the investigation of the diverse manifestations of consumer disempowerment associated with smart meters, an extensive search was conducted for peer-reviewed academic articles on smart meters in Europe published between 2010 and 2021 and incorporating qualitative findings. Qualitative studies, in which respondents can express their concerns, are indeed a relevant method of gathering information to understand the potential sources of consumer disempowerment.

The review highlighted four recurrent sources of consumer disempowerment in the context of smart meter, that are presented below.

3.1 Privacy Invasion and Information Panopticon

A recurring concern across European countries was the potential of smart meters and other smart home technologies to compromise households' security and invade their privacy (Balta-Ozkan et al. 2014; Nilsson et al. 2018). For example, respondents were worried about being "bombarded with different energy suppliers" trying to compete for their custom. Participants also expressed concerns over third parties knowing daily routines and occupancy, data falling into the wrong hands. In addition, consumers reported resisting smart meters because they viewed them as an extension of powerful companies into their private lives and the home domain. Respondents invoked the metaphorical term of "Big Brother" to express the fact of being watched by an unseen and intrusive presence. As pointed out by Savirimuthu (2013), smart meters are interpreted by consumers as an "information panopticon", which gives government or corporate entities significant access to private consumer data, as a mechanism of power and a diagram of political technology.

3.2 Loss of Individual Autonomy

Issues of a lack of individual autonomy emerge as a key source of consumer disempowerment regarding smart meters (Buchanan et al. 2016). Consumers express their uneasiness about the idea of energy suppliers managing their energy consumption for them. They have the feeling that smart meters will let their electricity company control their energy use without their full consent. Such findings replicate previous research which has found that consumers voice concerns about losing control when presented with the concepts of remote disconnection or smart appliances (Fell et al. 2014). Such objections regarding a potential loss of individual autonomy may also stem from consumers' thoughts that their energy supplier will at the end prevent them from using energy as and when they wanted, this leading to a decline in comfort and the disruption of well-established household routines.

3.3 Rejection of Additional Measures to Reduce Energy Consumption at the Household Level

A significant source of disempowerment associated with the smart meter is the perception by households that there is no way to save more energy without significantly reducing their comfort and freedom to act within their homes (Buchanan et al. 2016). In this case, energy consumption is perceived as "already as low as possible", while daily activities such as cooking, laundry and cleaning are considered as basic needs and therefore "out of control". As a result, feedback on energy consumption is perceived as generating anxiety in households. Thus, consumers fear being constantly exposed to their own consumption, and being reminded of the environmental and financial impacts of this consumption, while feeling that they have no control over it.

3.4 Mistrust of Energy Suppliers and Price Vulnerability

Existing qualitative research about smart meters consistently found that consumers mistrust of energy suppliers acts as a barrier to consumers' willingness to adopt demand-responsive enabling technologies (e.g., Fell et al. 2014; Buchanan et al. 2016; Sovacool et al. 2017). Consumers seem to find it difficult to comprehend how an energy provider will profit if they are encouraging consumers to use less energy. In a similar way, they question the reasoning behind the financing of the implementation of smart meters, wondering which parties are likely to benefit from it. Generally, consumers feel that they are unlikely to be the ones to benefit from the smart metering initiatives, thus formulating theories about how the energy companies will ensure that they will ultimately benefit from smart meters. Consumers fear that energy suppliers will charge them for the cost of the new meters in their bills, profiting from the interest on consumers' energy savings. A similar concern is that smart meters will be used to shift responsibility for augmented energy bills onto households on the grounds that smart meters enable them to make 'better' energy management decisions, thus negating supplier obligations to ensure that householder bills are kept as low as possible. Additionally, they suspect that energy suppliers will implement new peak-related tariffs that will cause energy to be more expensive when consumers will most likely need it. In the digital age, companies tend to charge customers individual prices determined based on their collected data, which is referred to as personalized pricing (Seele et al. 2021), as a method to generate economic returns. As consumers are becoming increasingly aware of such company pricing practices, they are likely to associate smart metering with potential energy suppliers' pricing manipulation through data collection and analysis.

In line with these results, a recent study examining the structure of social representations concerning smart meters in France showed ambivalent themes of customer psychological empowerment, which reflect the sources of disempowerment mentioned above (Schweitzer 2021). This study is in line with Moscovici's (1961) social representations approach, which focuses on the social context in which consumers learn, perceive their environment, and develop points of view, and thus refers to socially constructed and shared knowledge, or the common sense of existing objects. The participants' central representations of smart meters revealed threats to their privacy, linked to the possibility for energy suppliers to disclose their private data. Secondly, consumers expressed a major concern about their autonomy, arising from the introduction of remote monitoring with the new metering device. Respondents' central representations also reflected perceptions of political subjugation and the weight of organizational injunctions to optimize their electricity consumption, while lacking information on how to do so. These injunctions were seen as likely to reduce their freedom of action within the household and cause anxiety. Finally, with a lesser impact, perceptions of an unfair exchange with energy suppliers are found among the peripheral social representations of smart meters, confirming the fourth theme of disempowerment of consumers outlined above.

4 Recommendations on How to Improve Consumer Empowerment in the Use of Smart Meters

The consistent finding that households report a diversity of psychological disempowerment manifestations regarding the use of smart meters should be interpreted neither as a fixed state, nor as a reason to give up on attempting to improve their acceptance. Instead, we suggest that smart meters currently constitute a limited tool for demand-side management, because consumer psychological empowerment was neglected until now.

First, to improve technology acceptance when consumers perceive privacy risks of using a device, extended research have suggested for brands to be much more focused on customers' positive outcomes in their communication campaigns. Indeed, researchers have outlined the trade-off operated by consumers which consists of comparing the benefits with the risks associated to the use of the artefact. Following this stream of research, we suggest energy suppliers to focus on the positive economical and ecological impact of using smart meters in their communication. Further, they should adapt their message depending on the main motivation of their target, which can be economic or environmental, and show how smart meters allow consumers to make informed decisions to reduce their energy consumption.

Second, while remote monitoring with smart meters can be perceived as a threat on their autonomy for several consumers, energy suppliers might also outline how smart meters can be involved in the process of better organizing their everyday life. Lately, an increasing number of brands have allowed consumers to choose a preferred schedule such as for the withdrawal of their online shopping, or for the home delivery of their online purchases. We can think of smart meters as a tool allowing consumers to schedule when their energy supplier remotely manipulates energy installations in their home.

Third, our findings underline the challenge to reduce anxiety stemming from the perception of having few possibilities to reduce energy consumption (Hargreaves et al. 2010). At the same time, anxiety is reinforced by perceptions of being pressured by companies, and more broadly macrostructures, to reduce energy consumption. Following Shove et al. (2012) which argued that individual practices are being constantly renegotiated, and that feedback plays an important role in the "persistence, transformation and decay of one practice (Shove et al. 2012, p. 99)", we suggest energy suppliers to improve the learning process of the smart meter feedback system. Supporting consumers' learning with benevolence is likely to reduce anxiety related to energy consumption reduction. More specifically, energy suppliers should train customers to use the feedback platform through technical manuals, remote employee support, as well as interactions with other users which constitutes a source of social support. Energy companies should also focus on designing smart feedback systems that are user-friendly and easy to use. Further, in a recreational approach, they could try to make energy consumption reduction a more playful activity, and implement challenges with opportunities to earn badges.

Finally, our results highlight consumers' mistrust of energy suppliers, more specifically concerning their intentions to increase energy tariffs in a near future. Indeed, individuals find it difficult to comprehend how energy providers can financially benefit from the roll out of smart meters, which encourage energy reduction consumption. We consider that a way to restore consumer trust might be for energy brands, to clarify that smart meters were implemented in a context of global energy transition agreements. This means energy suppliers should use pervasive communication on the purpose of smart meters' roll out, which constitute an attempt to involve consumers in the reduction of energy consumption worldwide. If consumers look at energy suppliers as actors of a global ecological program and not acting for their own financial benefits, they would be reassured and not infer price manipulation intentions about energy suppliers involved in the implementation of smart meters. Further, energy suppliers could also clarify the role of smart meters in the development of a "smart grid", which aims to improve the distribution of energy for consumers. Table 1 summarizes our recommendations related to each disempowerment source identified.

5 Conclusion

Investigating the underlying factors sustaining customer resistance to smart meters, this chapter outlines four sources of consumer psychological disempowerment from previous consumer qualitative research. Specifically, privacy invasion, loss of individual autonomy, pressure on energy reduction measures and mistrust of energy

Table 1 Recommendations for energy suppliers to reduce consumer disempowerment sources in the use of smart meters

Consumer disempowerment sources	Recommendations for energy brands
Privacy invasion perceptions and information panopticon	• Build communication campaigns focusing on the positive economical and ecological impact of using smart meters • Adapt the message depending on the main motivation of their target, which can be economic or environmental, and show how smart meters allow consumers to make informed decisions to reduce their consumption
Loss of individual autonomy	• Outline how smart meters allow consumers to schedule when their energy supplier remotely manipulates their installations, and are thus involved in the process of better organizing their everyday life
Pressure for additional energy reduction measures at household level	• Improve the learning process of the smart meter feedback system through technical manuals, remote employee support as well as interactions with other users as a source of social support • Make energy consumption reduction a more playful activity, and implement challenges with opportunities to earn badges for instance
Mistrust of energy suppliers and price vulnerability	• Restore consumers trust using pervasive communication clarifying that smart meters are implemented in a context of global energy transition agreements, and not for energy suppliers' own financial benefits

suppliers pricing strategies intentions explain the gap between companies' expectations and actual acceptance of smart meters. Rather than seeing these disempowerment manifestations as fundamental limitations of all smart meters for energy suppliers, we have formulated recommendations for marketing managers involved in the roll out of smart meters. We suggest improving their communication campaigns by focusing on consumers' benefits, designing user-friendly and playful smart feedback interfaces, and clarifying the underlying motives of smart meters roll out, which aims to reduce global energy consumption. All these measures should help reducing disempowerment factors associated to smart meter use.

While our chapter focuses on disempowering factors, this does not mean that smart meters do not convey positive psychological perceptions of empowerment. These could, through the prism of the customer's psychological compensation mechanisms, mitigate the negative effects of disempowering representations. Therefore, the sources of psychological empowerment of consumers should be further investigated.

References

Balta-Ozkan N, Boteler B, Amerighi O (2014) European smart home market development: public views on technical and economic aspects across the United Kingdom, Germany and Italy. Energy Res Soc Sci 3:65–77

Batalla-Bejerano J, Trujillo-Baute E, Villa-Arrieta M (2020) Smart meters and consumer behaviour: insights from the empirical literature. Energy Policy 144(C)

Buchanan K, Banks N, Preston I, Russo R (2016) The British public's perception of the UK smart metering initiative: threats and opportunities. Energy Policy 91:87–97

Chamaret C, Steyer V, Mayer JC (2020) "Hands off my meter!" when municipalities resist smart meters: linking arguments and degrees of resistance. Energy Policy 144:111556

Darby S (2006) The effectiveness of feedback on energy consumption: a review for DEFRA of the literature on metering, billing and direct displays. Environmental Change Institute, University of Oxford, Oxford

Delmas MA, Fischlein M, Asensio OI (2013) Information strategies and energy conservation behavior: a meta-analysis of experimental studies from 1975 to 2012. Energy Policy 61:729–739

European Commission (2020) Directorate-general for energy. In: Alaton C, Tounquet F (eds) Benchmarking smart metering deployment in the EU-28: final report. Publications Office (2020). https://data.europa.eu/doi/10.2833/49207

Fell MJ, Shipworth D, Huebner GM, Elwell CA (2014) Exploring perceived control in domestic electricity demand-side response. Technol Anal Strateg Manag 26(10):1118–1130

Frederiks ER, Stenner K, Hobman EV (2015) Household energy use: applying behavioural economics to understand consumer decision-making and behaviour. Renew Sustain Energy Rev 41:1385–1394

Hargreaves T, Nye M, Burgess J (2010) Making energy visible: a qualitative field study of how householders interact with feedback from smart energy monitors. Energy Policy 38

Mela H, Peltomaa J, Salo M, Makinen K, Hilden M (2018) Framing smart meter feedback in relation to practice theory. Sustainability 10(10):3553

Menon S (2001) Employee empowerment: an integrative psychological approach. Appl Psychol 50(1):153–180

Moscovici (1961) La psychanalyse, son image et son public. Paris, PUF

Nilsson A, Wester M, Lazarevic D, Brandt N (2018) Smart homes, home energy management systems and real-time feedback: lessons for influencing household energy consumption from a Swedish field study. Energy Build 179:15–25

Parish B, Heptonstall P, Gross R, Sovacool BK (2020) A systematic review of motivations, enablers and barriers for consumer engagement with residential demand response. Energy Policy 138

Savirimuthu J (2013) Smart meters and the information panopticon: beyond the rhetoric of compliance. Int Rev Law Comput Technol 27(1–2):161–186

Schweitzer V (2021) How a smart technology imposed by an organization can arouse ambivalent psychological empowerment: an approach leveraging the theory of social representations. J Organ Psychol 21(3):143–164

Schweitzer V, Simon F (2021) Self-construals as the locus of paradoxical consumer empowerment in self-service retail technology environments. J Bus Res 126:291–306

Seele P et al (2021) Mapping the ethicality of algorithmic pricing: a review of dynamic and personalized pricing. J Bus Ethics 170(4):697–719

Shove E, Pantzar M, Watson M (2012) The dynamics of social practice: everyday life and how it changes. Sage, London

Sovacool B, Burke M, Baker L, Kotikalapudi C, Wlokas H (2017) New frontiers and conceptual frameworks for energy justice. Energy Policy 105

Spreitzer GM (1995) Psychological empowerment in the workplace: dimensions, measurement, and validation. Acad Manag J 38(5):1442–1465

Smart Meters Improved by NILM

Daniel Weißhaar, Pirmin Held, Dirk Benyoucef, Djaffar Ould Abdeslam, Patrice Wira, and Jean Merckle

1 Introduction

This chapter introduces the technology Non-Intrusive Load Monitoring, a method for detecting individual devices from an overall signal. Non-Intrusive Load Monitoring is the research area and technology behind the third word in Smart Meter Inclusive. Using a smart meter as a basis and recognizing devices from the power profile is not a new idea but is now a common practice in Non-Intrusive Load Monitoring. However, the approach to creating such a measurement system that classifies appliances in real-time and visualizes the results directly on the same hardware has not been existing yet. Smart Meter Inclusive wants to leave the data where it originates, namely with the customer. This book chapter provides a general overview of non-intrusive load monitoring to be able to understand the basics and approaches for such a Smart Meter Inclusive.

2 Efficient Energy Monitoring Through Non-intrusive Load Monitoring

One of the most important issues of our time is environmental protection. Everyone owns more and more electronic appliances; politicians rely on electric vehicles; energy consumption continues to rise. Of course, this is compensated for by ever more efficient devices. Washing machines and dryers use significantly less energy today than ten years ago. However, this energy efficiency alone is no longer sufficient.

D. Weißhaar (✉) · P. Held · D. Benyoucef
University of Furtwangen, Furtwangen, Germany
e-mail: daniel.weisshaar@uha.fr

D. Weißhaar · D. Ould Abdeslam · P. Wira · J. Merckle
University of Haute Alsace, Mulhouse, France

D. Ould Abdeslam (ed.), *Smart Meters*, Lecture Notes in Energy 97,
https://doi.org/10.1007/978-3-031-27556-2_3

One approach to address this seemingly endless increase in energy demand is to use consumers more efficiently (Hoyo-Montaño et al. 2016; Luca et al. 2015). A possible idea of the implementation is to make the customer's consumption more transparent to be able to show him the resulting potential savings. This transparency opens a new problem. How do you measure the consumption of individual devices? And is it a sensible concept to install more and more power meters? Each of these devices consumes energy itself. But in order to identify all consumers with this concept, each consumer must be equipped with its own electricity meter. Power meters are not cheap. In order to evaluate the data, they have to be synchronized first. Some devices may be installed permanently and inaccessible. Therefore, it is difficult to add a power meter after the fact. And finally, power meters themselves are new consumers.

Developments for more efficient consumption monitoring in electronic networks began at MIT in the USA in the 1990s (Hart 1989). A basic physical principle is that the power of consumers that are running at the same time can be superimposed. Hart postulated at that time that electronic consumers could be divided into different main categories. He also found through recording that various devices have very individual behavior. Hart's conclusion was that a kind of reengineering could take place here. So instead of installing an electricity meter in front of each individual consumer, only one meter is required at the entrance to the electronic network. This central meter allows the measurement of the superimposed total consumption of all devices. These individual devices are calculated out of the total consumption with the help of various signal processing and artificial intelligence methods. A kind of inverse superposition takes place.

Another motivation for monitoring and detecting consumers can be found under the keyword Ambient Assisted Living (AAL) (Bucci et al. 2021; Ruano et al. 2019; Klein et al. 2013). Here, Non-Intrusive Load Monitoring (NILM) can help monitoring the health status of older people. For the most part, AAL relies on sensors of all kinds, from direct vital signs sensors in smartwatches to indirectly incorporated accelerometers in smartphones. Usually, these sensors have to be carried actively on the body and therefore require a certain tolerance of the wearer. NILM enables an indirect insight into the everyday behavior via appliance recognition without intervening on the freedom of the person or forcing a change in behavior. It is not necessary to procure several expensive sensors for this, but every household appliance automatically assumes the role of such a sensor.

2.1 Load Disaggregation and Other Terms

When Hart began his work in this area of research, he coined the name Nonintrusive Appliance Load Monitoring (NIALM) from the title of his publication of the same name (Hart 1992). Over the years, other terms and forms of writing derived from them have been established, but they can all be used synonymously. The most common representatives of these are Nonintrusive as well as Non-Intrusive Load

Monitoring (NILM) or Nonintrusive as well as Non-Intrusive Appliance Load Monitoring (NIALM). Another abbreviation is NALM, where the non-intrusive is abbreviated to just one letter. Energy or load disaggregation is also quite common. Disaggregation derives from the idea that the aggregated performance i.e. the resulting total performance of all devices, is measured at a central measuring point. Mathematically, the aggregated power can be expressed as Eq. 1.

$$P_{tot}(t) := \sum_{i=1} (P_i(t))^N + e(t) \tag{1}$$

The N individual loads are described by $P_i(t)$, and an additional disturbance term $e(t)$ is added, which describes both the noise and possibly unidentifiable loads. So, disaggregation is the inversion of this aggregated signal into its consumer signals.

In this book, we mainly use the notation most commonly used today, Non-Intrusive Load Monitoring, and the abbreviation NILM.

2.2 *Intrusive Versus Non-intrusive*

One point that inevitably gets stuck when examining the name of the research area is non-intrusive. So the question arises of what intrusive and non-intrusive mean. The two terms refer to the measuring principle. Intrusive means that a measuring device is on the electronic network, while non-intrusive represents a black box measuring method.

The simplest example of an intrusive measurement is through a plug-in power meter. This measuring device is attached between the actual socket and the appliance. It also creates a load drop itself, although usually relatively small. However, it requires an intervention in the network structure since it has to be connected in between. In the case of a non-intrusive measurement, measurements are taken outside of the network to be analyzed. The current measuring methods for NILM use the physical principle that every current also induces a magnetic field. This means that the sensors can easily be retrofitted around the individual current phases.

Both measurement methods, intrusive and non-intrusive, can be found. They are summarized under the term Appliance Load Monitoring (AML) (Hart 1992). Intrusive monitoring of appliances is the classic example that everyone immediately has in mind. Each device to be monitored has its power meter. The problem with this measurement method is that the power meters are expensive to purchase and install. Furthermore, the recorded data must be combined to be able to evaluate them. On the other hand, there is the non-contact monitoring of devices by measuring at a central point in the network. The non-intrusive method is significantly cheaper to install and easier to retrofit. The problem with this approach is that only the aggregated signal is available due to the central measurement. The load disaggregation must therefore be implemented in software using various algorithms from signal processing and artificial intelligence.

2.3 Process Chain of Non-intrusive Load Monitoring

Over the years, a standard procedure for the implementation of NILM has been established, which can be found again and again in this or a slightly modified form. These individual components of the processing chain ensure that the complex NILM problem is broken down into more manageable sub-problems. The structure for event-driven NILM approaches is shown as an example.

- Data Acquisition (current and voltage)
- Preprocessing (filtering, but also conversion to common formats like P&Q, Harmonics, ...)
- Feature Extraction (steady-state features, transient-state features)
- Event Detection
- Classification
- Monitoring (depending on the objectives, can be approximation of consumption or detection of anomalies, ...).

The individual steps are described in more detail below.

Data Acquisition

Data acquisition involves measuring a signal. Current and voltage can be measured here, but also the power itself can form the input signal using a power meter. An important parameter related to data acquisition is the sampling rate. In this case, sampling below 1 Hz is referred to as a low sampling rate, while high-frequency sampling in the context of NILM is in the range of > 1 kHz to the MHz range (Zhuang et al. 2018).

Preprocessing

In the preprocessing stage, the recorded signals are subjected to an initial adjustment. If necessary, digital filtering can take place here. Signal conversions, such as power calculations, also count as preprocessing in this sense. The signal is converted into a form that can be used for event detection and feature extraction.

Feature Extraction

Depending on the objective, a feature is obtained from the preprocessed signal. This feature can later be used for classification. A wide variety of methods are used here, which can deliver one-to-multidimensional features. A simple example is the power values P&Q per period. Other examples are the harmonics or V-I-trajectories.

Another distinction in feature extraction is the question of the signal section to be used for extraction. There are Steady State Features (SSF) and Transient State Features (TSF). With the SSF, the signal change before and after an event is compared. The SSF is particularly suitable for low-dimensional features. TSFs, in turn, cover an entire signal section, from the beginning of the occurring signal change of an event to the transition when the signal again assumes a quasi-stable state. By considering this dynamic change, TSF delivers multidimensional features.

Event Detection

In event detection, an algorithm is applied to the recorded signal to detect when an appliance has changed its state. The bandwidth for such methods ranges from simple threshold detectors to complex methods such as wavelet transforms.

Classification

Once events are detected, they can be examined more closely. An attempt is made to identify the appliance causing the event. Any classification method from the field of machine learning can be used to solve this problem. Not included in the representation of this process chain is the fact that such a classification method must be learned beforehand.

Monitoring

In monitoring, the goal is crucial. Once it is clear which information the NILM processing chain should deliver, these results can be combined with the knowledge gained so far. The motivations for NILM listed at the beginning, energy reduction and ambient assisted living, alone indicate how broadly monitoring is to be understood.

2.4 Appliance Categories

From the outside, some appliances look similar, while others are fundamentally different. This can also be observed inside. While two different toasters have a simplified internal structure consisting of a heating element and are therefore similar, they can be distinguished from a fan, for example, by a completely different type of electronic consumer in the signal curve. Many different appliances have been analyzed in research on NILM. The division into four basic categories of electronic devices has become established in the literature on NILM (Abubakar et al. 2015; Hart 1992; Zeifman and Roth 2011; Zoha et al. 2012), which are presented in Table 1.

These appliances are further divided into event-based and eventless appliances. An event is a transition from one state to the next or the turning on or off. Event-based appliances include Type I and II devices. Eventless devices are Type III and IV. On the consumption side, changes in the status of the latter two device types cannot usually be recognized or clearly defined.

2.5 Event-Driven Versus Eventless Approach

There are two approaches to solving the problem of detecting devices within the framework of NILM. As already described, one way is to search for events in the incoming signal. This path requires a suitable event detection method, which must be adapted to the respective situation in the network. The approach of wanting to

Table 1 Appliance categories

Category	Description	Typical examples
Type I: on–off-devices	Simple devices that can only be switched on and off	Lamps, toasters
Type II: finite state machine	Devices that can assume a countable number of states and whose state changes usually follow a fixed program	Washing machines, dishwashers, coffee machines
Type III: various power devices	Devices that do not know any fixed states, the performance changes fluently	Drilling machine
Type IV: permanent consumer	Devices that are switched on permanently or for a very long time	Steady light, satellite receiver

recognize devices without an event usually amounts to an algorithm based on a Hidden Markov Model. Both approaches have their areas of application.

Event-Driven Approach

In the event-driven approach, an event detection method is used. It aims to identify events that are occurring as precisely as possible. Events are changes in the signal in the classic sense, which lead from one steady-state to another. How well such an algorithm can work depends on the device constellation within an electronic network. There are very power-intensive consumers, as well as small consumers. If there are only large consumers in a network that are to be recognized, there is a good chance for the algorithm to produce good results. It is the same in a network in which only small consumers appear. Two problematic quantities limit the result. On the one hand, there is the strength of the noise in the network; on the other hand, there is the power difference of the smallest event change. If the background noise is already greater than the smallest event, it is not detectable with certainty. The same problem also comes into play when large and small consumers are together network. Small events then threaten to be lost in the dynamic behavior of large devices.

There are different approaches to implementing event detection algorithms. These approaches range from threshold-based methods and statistical tests to neural networks (Held et al. 2018b; Lu and Li 2020; Wild et al. 2015; Yang et al. 2020).

The F_1 score can be used to validate the quality of such an event detection algorithm. This is based on the key figures of a binary classifier.

Another key figure is true negative, which is not determined in this specific problem.

The three values of Table 2 are used to calculate precision (Eq. 2) and recall (Eq. 3):

$$precision := \frac{TP}{TP + FP} \tag{2}$$

Table 2 Key figures of a binary classifier

True positive (TP)	Events detected by the event detector that were correct
False positive (FP)	Events detected by the event detector that were wrong
False negative (FN)	Events that were not recognized by the event detector

$$recall := \frac{TP}{TP + FN} \tag{3}$$

The value precision indicates how exactly all events that actually occurred were recognized. In contrast, recall expresses how high the correct detection was for all events detected by the event detector.

Both, precision and recall, give a value between 0 and 1 for the quality of the event detector. With the F_1 score, there is another value that combines these two in a common quality criterion. The F_1 score (Eq. 4) is the harmonic mean of recall and precision.

$$F_1 := 2 \cdot \frac{precision \cdot recall}{precision + recall} = \frac{TP}{TP + \frac{1}{2}(FP + FN)} \tag{4}$$

With the help of the F_1 score, the quality of the event detector can be expressed in just one number, which is advantageous for comparability and optimization.

Eventless Approach

With the eventless approach, a different strategy is followed than paying attention to individual abrupt changes in the signal. Algorithms based on factorial Hidden Markov Models (fHMM) (Ghahramani and Jordan 1995) are used here. These models use previously determined probabilities to estimate which devices are involved in the current overall signal and in what form.

The Hidden Markov Models (HMM), on which the fHMM is based, are state machines that are not directly accessible. However, it is known from which state S_i into which other states S_j can go (transition) and with what probability this occurs. This results in the transition matrix A. The observable outputs of the HMM Y_t at the time t are called emissions, whereby the transition from the respective state S to the emission Y is described by the emission matrix B. Y_t shows an observation of a predefined value set $O := [O_1; O_2; \ldots; O_M]$. Since the states S_i cannot be observed directly, they are referred to as hidden states. An example of HMM can be seen in Fig. 1.

The fHMM in turn is a combination of many such HMMs. It is assumed that the individual HMMs are independent of each other. This results in an observable emission Y_t for a time t and a resulting state vector S_t with $S := [S^{(1)}, S^{(2)}, \ldots, S^{(N)}]$. Transferred to NILM, each of the N devices is in a state $S_t^{(i)}$ at any point in time

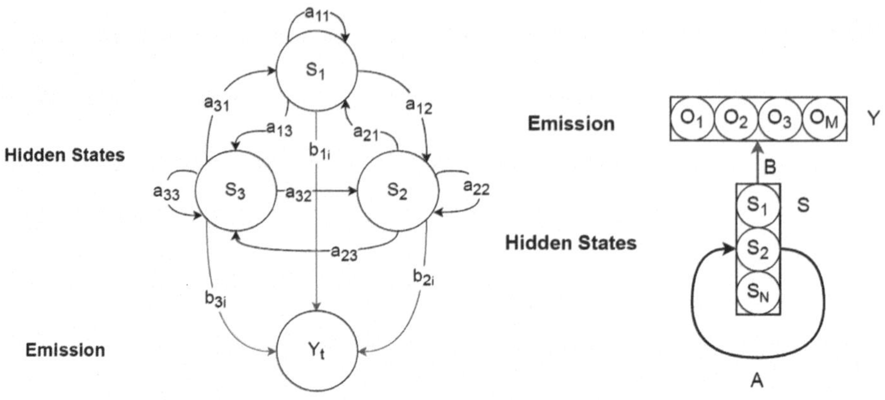

(a) Example of a Hidden Markov Model (b) Simplified structure of a Hidden Markov Model

Fig. 1 Hidden Markov Model

and Y_t is the metric to be observed, e.g. the overall performance of the network. An exemplary representation can be seen in Fig. 2.

An electronic network in which all consumers are known is required. Consumer status may vary. A separate HMM describes which states a consumer can assume. For this description, each device must first be examined for the number of possible states. This is followed by the development of a suitable statistical model for each device. In the fHMM, the individual HMMs are then combined into a common model.

The problems arising from the use of fHMMs are the creation of the individual HMMs for the devices. For this purpose, the transition matrix and the emission matrix are calculated in a training phase. If only little is known about the devices, state estimation methods can be used (Egarter et al. 2015). The identification of the individual states within the framework of the classification is often solved with the Viterbi algorithm (Yang et al. 2021).

Use of Both Approaches

The eventless approach is particularly suitable for devices with very slow state changes. In addition, only a low sampling frequency is required for this, since the state changes do not have to be specifically detected. The eventless approach relies on probability to find the combination of devices that cause the current overall performance. However, if several devices of the same or similar type are to be kept apart, the fHMM approach is not suitable for keeping them apart. In this case, an event-driven approach with a high sampling rate is advantageous because it separates event detection from classification. A special algorithm can thus deal with the issue of precise device recognition, which in this case will lead to more precise results.

The eventless approach is suitable for appliances of types I, II, and IV. Since devices of type IV have no or only very rare events, it is difficult to recognize them with the event detection method. Additional boundary conditions would have to be inserted for this. With the eventless approach, appliances of type IV can simply be

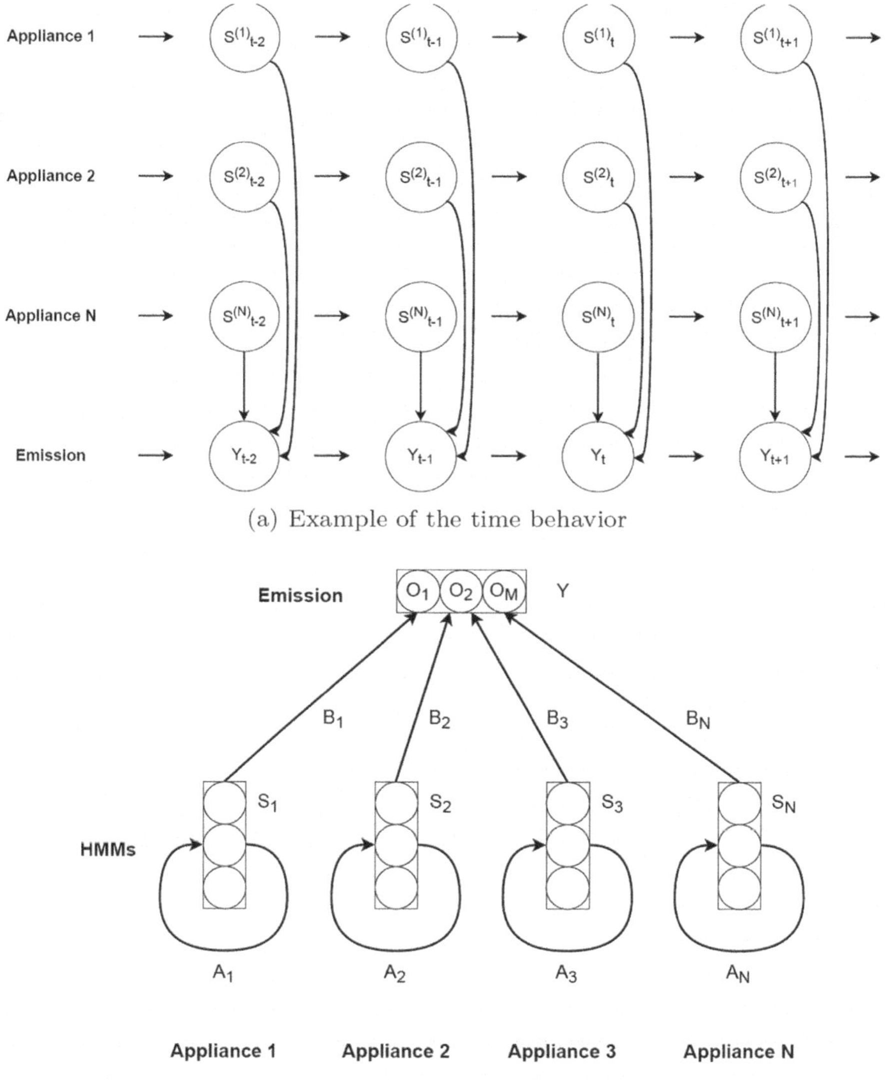

(a) Example of the time behavior

(b) Simplified structural description

Fig. 2 Factorial Hidden Markov Model

taken into account, like all other devices. However, in the form presented here, both approaches have problems with type III. Varying performances cannot be assigned to a state with certainty and are probably only partially recognizable with the eventless approach, especially in a dynamic phase.

3 Measurement Systems

In many works on NILM, Smart Meter recordings are used as the data source. Again, a lot of the research work is based on recorded data sets and focuses on the development of algorithms. In other research work, special hardware solutions are developed in-house. This section deals with the different sensors and measurement hardware. In addition, a few existing and publicly accessible datasets are presented.

3.1 Sensors for Non-intrusive Load Monitoring

Non-intrusive load monitoring begins with a measurement signal. This signal is measured by sensors. Classically, NILM measuring devices rely on a contactless measuring method. Various current measurement methods are presented below, which can be found in smart meters and NILM measuring devices.

Shunt Resistor

The simplest measuring principle to measure the electric current is a shunt resistor. The shunt resistor is installed in series in the current path. The electrical current can then be measured as voltage V_S through the contact points using Ohm's law and the defined resistance R_S. The basic measuring principle is shown in Fig. 3.

For the measurement, the existing network must be interrupted at one point in order to be able to use the resistor. It should be noted that R_S must be as small as possible here so that the influence on the existing network is as small as possible and the total resistance can be neglected alongside the actual consumers R_L. Then Eq. 5 applies to the current measurement accordingly.

$$i(t) = \frac{V_S(t)}{R_S} \tag{5}$$

The current is calculated with sufficient accuracy using an amplifier circuit and an evaluation circuit. This measuring principle can be found in part in smart meters. It works with both direct current and alternating current.

Current Transformer

The folding coil current transformers are based on a measuring principle of induction of current-carrying conductors. Two coils are wound on a ferromagnetic core. The primary side is the input side with the current-carrying conductor that is to be measured. On the output side, the secondary side, there is a defined number of turns and the outlets of the measuring line. In Fig. 4 this measuring principle is shown schematically.

If the number of turns N_1 and N_2 is known, the unknown primary current I_P can be measured via the secondary current I_S. Then Eq. 6 applies.

Fig. 3 Current measurement principle with a shunt resistor

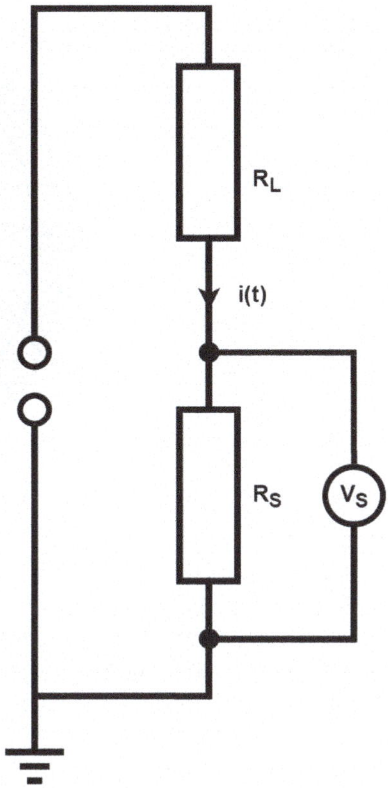

Fig. 4 Current measurement principle with a current transformer

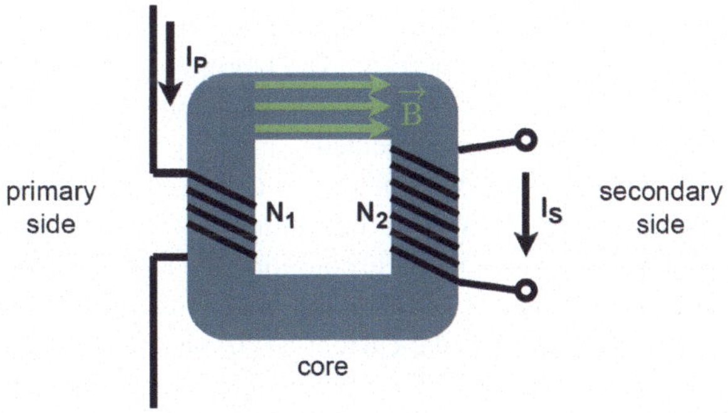

$$I_P = \frac{N_2}{N_1} \cdot I_S \tag{6}$$

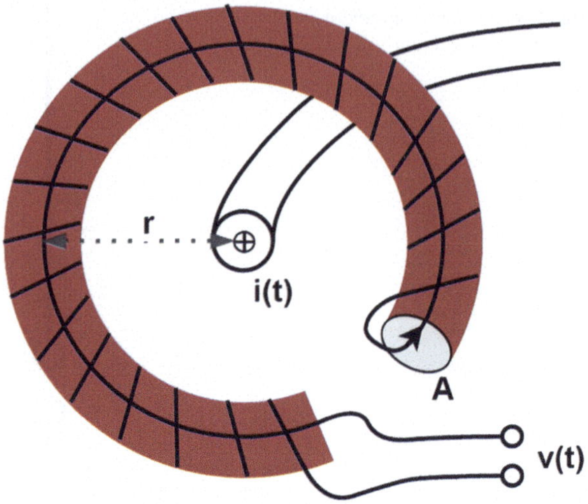

Fig. 5 Current measurement principle with a Rogowski coil

The electrical input current can be easily determined by means of an evaluation circuit on the secondary side. The relationship between the input and output sides is usually given in the datasheet about the current ranges. With some current transformers, such as a folding transformer, the core can be opened and thus fitted around a conductor. This conductor is then only passed through the core once and thus has the number of turns $N_1 = 1$.

Rogowski Coil

The Rogowski coil can also be used to measure current. This is an annular air-core coil without a metal core, as shown in Fig. 5. The coil is passed through a ring and wound back around the ring. The beginning and the end of the ring are not tightly closed. If the ring ends are placed close enough together, the inhomogeneity of the magnetic field can be neglected.

The Rogowski coil is placed as a ring around a current-carrying conductor. The current-carrying conductor has a magnetic field in which the Rogowski coil is now located. Due to the magnetic coupling, the Rogowski coil experiences self-induction. A voltage can be measured at the open ends of the Rogowski coil. This voltage can be directly related to the current $i(t)$ to be determined by means of an amplification and evaluation circuit. The measuring principle using the Rogowski coil only works with alternating current, since the change in current induces the voltage.

Hall Sensors

Another way to measure the current without contact is with a Hall sensor. Here, the Hall effect is used. The basic measurement principle is shown in Fig. 6.

The current-carrying conductor to be examined with the current I_M is led through a ferromagnetic core. This core has an air gap in which the Hall probe is located. The flow of current I_M generates a magnetic field in the core, which penetrates

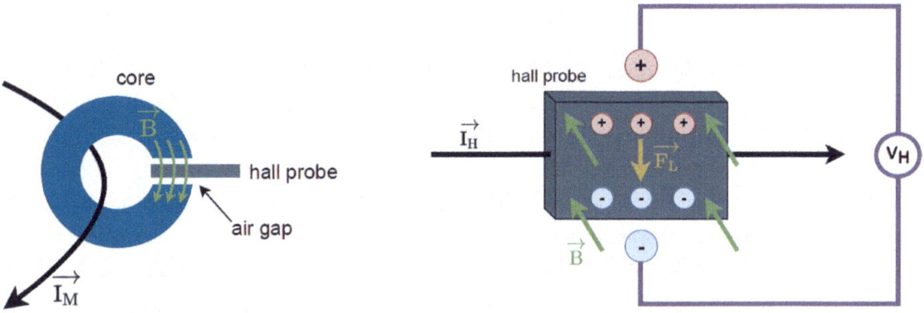

(a) Principle of the hall sensor (b) Hall effect on the hall probe

Fig. 6 Current measurement principle with a Hall sensor

vertically through the Hall probe. A known current I_H flows through the Hall probe itself. The Hall effect generates a Lorentz force F_L. This Lorentz force separates the charge carriers in the Hall probe. This creates an electrical field proportional to the magnetic field, which can be measured as a Hall voltage V_H. Because the dimensions of the core, the air gap, the material of the Hall probe, and the current through the Hall probe are known, the current I_M can be calculated using the measured voltage V_H.

Comparison

Each measurement method has its advantages and disadvantages. Shunt resistors, for example, are very cheap and have a simple circuit design. However, they can only be introduced into the current path and cannot be attached without contact with a conductor. The various measuring methods also work with different degrees of accuracy. A study from 2016 (Leferink et al. 2016) was able to show that, depending on the type and behavior of the electronic consumers, various measurement methods sometimes show massive differences in the measurement results. The Rogowski coils, in particular, sometimes showed considerable deviations when there were rapid load changes. Again, this effect turned out to be a positive property for event detection and classification in the context of NILM in Held et al. (2020), especially when it comes to distinguishing between very similar devices.

3.2 *Measurement Hardware*

Developing your hardware is a very time-consuming process and is therefore not always the first choice. Depending on the goal of your research, the focus can only be on developing algorithms or on setting up an entire NILM measurement system. Many therefore work with public data sets, which meanwhile abound. Works that rely on in-house measurements either try to use existing solutions or develop specific hardware to meet their own needs.

Smart Meter Based Data Acquisition

Developing your hardware is often expensive, time-consuming, and may not even be the focus of research work. A direct connection to a Smart Meter is used in many publications. From a pragmatic point of view, one day every household will be equipped with a smart meter. Many Smart Meters have different interfaces. This means that the measurement data can be accessed directly. One advantage is that there is no need to develop and calibrate your measuring unit. A disadvantage may be the low sampling rate. While the energy companies receive the measured values at intervals of several minutes to one hour (Adabi et al. 2016; Liang et al. 2019), the measured value can be queried directly on the device via interfaces such as Modbus, sometimes with a frequency of a few minutes to around 1 Hz (Bousbiat et al. 2020; Raiker et al. 2018). Furthermore, the output format cannot be freely selected. A Smart Meter provides performance values, not voltage and current. Accordingly, the possibilities in terms of feature extraction are limited. However, for use cases in which such low sampling and the power values are sufficient, Smart Meters offer a simple and stable option as a basis for data acquisition. In some cases, the smart meters are supplemented by other elements such as openHAB for data logging and evaluation (Bousbiat et al. 2020). The development of a NILM measuring system based on Smart Meters has a few limitations as well as decisive advantages. The hardware already exists, no calibration of the sensors is required, and the downstream hardware or software can be kept very simple. It is therefore also understandable why this path is followed in many research projects.

Self-developed Hardware for Data Acquisition

Smart Meters are a real alternative for data collection. However, if special questions require a high sampling rate, for example, or if the goal in an industrial context is to take a closer look at several machine systems, Smart Meter-based hardware approaches may no longer be so suitable. There are attempts to carry out a disaggregation with low sampling rates (Liang et al. 2019). However, it could also be shown that different devices cannot be distinguished with slowly sampled signals, while they can be distinguished without any problems with a higher sampling rate (Adabi et al. 2016). Furthermore, the features to be used have a significant impact on which sampling frequency is required (Dinesh et al. 2016; Zeifman and Roth 2011). In the case of in-house developments, a distinction can be made between FPGA-based solutions (Barbero et al. 2020; Cardenas et al. 2016; Trung et al. 2012) and those with microcontrollers (Shiddieqy et al. 2021; Yaemprayoon et al. 2016). In addition to the goal of developing a real-time NILM solution, there are also pure data loggers (Kolter and Johnson 2011). The data loggers in particular are needed to be able to record new data sets for algorithm development. Here, the disaggregation takes place separately from the hardware.

3.3 Public NILM Datasets

Developing your hardware is a separate topic that can take a lot of time and effort. Since there is no development hardware for NILM to buy, you have to develop it yourself. Using Smart Meters, there is already an almost generic way to produce your measurement data, but setting up your measurement scenario is also a very time-consuming undertaking. In the meantime, there are many different data sets in the research field, which cover a wide variety of goals. Some of them were developed in the laboratory, others are recordings from real households. The data sets differ, among other things, in the sampling frequency, the signal form, the length of the sequence, and the devices used. Pereira and Nunes (2018) provide a large overview of many data sets. A distinction can also be made between datasets recorded in households and datasets generated under laboratory conditions. Representatives for household datasets are REDD (Kolter and Johnson 2011), BLUED (Anderson et al. 2012), UK-DALE (Kelly and Knottenbelt 2015) and ECO (Beckel et al. 2014). Representatives for laboratory datasets are WHITED (Kahl et al. 2016), COOLL (Picon et al. 2016), PLAID (Gao et al. 2014). A problem with existing datasets was that often only single device measurements or aggregated measurements are available in the datasets. The dataset HELD1 (Held et al. 2018a), which was also generated under laboratory conditions, was developed to combine training and test sequences in a common dataset. Another particular dataset is HELD2 (Weißhaar et al. 2020). Only the individual measurements have been included here. The aggregated datasets were generated synthetically from the individual measurements under defined conditions. HELD2 is the first simulation data set for NILM.

4 Feature Extraction for the Appliance Classification

Good device recognition results can only be achieved later with well-prepared features. Therefore, feature extraction with its many possibilities is a topic that should not be neglected.

4.1 Steady State and Transient State

The goal of NILM is to detect devices and their states in a current or power signal. Various more general information can be recognized in the signals. In the event-based approach, two phases in the signal can be distinguished concerning a device, the steady-state, and the transient state.

The steady-state describes the state in which the signal behaves quasi-statically, i.e. does not experience any state change. Applied to a simple device, this is either the on or off state. The transient state designates the period in which the signal changes before it has passed from one steady-state to the other. The concepts of steady-state and transient state are elementary for event-based feature extraction since the entire signal is not processed here, but event-centric sections are processed.

4.2 Extraction of Features in NILM

Hart started with admittance as a signal form (Hart 1992), but has already described several other suitable signal forms that can be used for processing in NILM. With the features themselves, a basic distinction is made between steady-state features and transient state features. Steady State Features are formed by comparing two Steady States. In the simplest case, the signal curve is subtracted from one another over a defined time window before a transient state with the same time window after the transient state. The difference then forms a possible steady-state feature. The transient features simply use the temporal signal range during the transient state.

The literature provides many different options for choosing the appropriate feature. Zhang and Zhu (2019) compared various steady-state features and transient state features. The active power P and the reactive power Q are given here as the simplest form, which can be found in numerous publications. A promising feature is the V-I-trajectory over a signal period. It provides a strong separability of devices. Other options are the harmonics of the current signal, which can be calculated using Fast Fourier Transform, and the waveform of the current signal itself. In the transient state features area, the instantaneous power and the instantaneous current waveform are listed as possible options. The S-Transform (Martins et al. 2012) offers another transient state feature. In the area of the steady-state features, there are also the wavelet transform (Zoha et al. 2012) and eigenvalues (Liang et al. 2010).

In addition to the classic features, which are ultimately based on the current and voltage signal, there are further investigations that involve additional sensors. Here, for example, the temperature (Morán et al. 2020) can be found as an additional signal. Light intensity, acceleration sensors, acoustic sensors, and other environmental sensors are described in Bergés et al. (2010) as possible additions.

For the selection of a suitable feature or feature set, the question remains whether the computing intensity plays a role, what data is available, whether the hardware can be expanded, whether additional sensors can be installed, and how these different sources can be combined. In general, only the steady-state features remain when using low sampling. The transient states are often short and meaningful information content requires a high sampling rate. This in turn means higher demands on the hardware, because high sampling rates also mean more data to be processed. In many different works, the classic features have given good results. Depending on the question, other sensors could provide useful support.

5 Frequency Invariant Transformation of Periodic Signals

Frequency Invariant Transformation of Periodic Signals (FIT-PS) (Held et al. 2016, 2019a) is an algorithm developed for NILM. FIT-PS can be applied to discretely sampled periodic signals. The idea behind FIT-PS is to generate a multi-dimensional signal out of a one-dimensional signal. This makes it possible to detect changes in the signal at specific points in time over a period of time. The motivation for the FIT-PS transform is that sampled voltage signals are always subject to a certain scatter in the period duration since the mains frequency f_0 is not constant. As a result, the number of sampling points per period is not always the same. This also results in a slight difference within a period from other periods at the time the sample was drawn. So, the signal has a certain frequency dependency. Using FIT-PS this frequency dependence is eliminated by interpolation and each period has a constant number of sampling points.

The FIT-PS transform can be described as follows:

$$FITPS : R^L \rightarrow R^{K \cdot N}$$

Here, L represents the length of the discretely sampled signal. The parameter K represents the number of periods of the transformed signal, and the parameter N represents the number of sampling points per period in the transformed signal. The degree of freedom of the parameter n makes it possible to choose the dimensionality of the transformed signal yourself within certain limits. Only the sampling rate of the original signal defines the limit of how large N can be chosen considering the Nyquist–Shannon sampling theorem.

The FIT-PS transform according to the algorithm is performed in several consecutive steps. First, a resampling takes place, in which the original signal is gradually converted into a signal with newly calculated sampling points Eq. 7.

$$\text{Resampling} : S_{org} \rightarrow S_R \tag{7}$$

First, the trigger signal used to determine the period changes is defined. With NILM, the voltage is selected as the trigger signal. In a normal power grid, it is assumed that the voltage signal has a sinusoidal curve and that there is a clear point in time for the period change. First, the time stamp of the beginning of the period is determined by linear interpolation. The same happens with the time of the end of the period. $N - 1$ equidistantly calculated points in time are defined between these time stamps. This results in exactly N points in time for each period. At each point in time, the interpolated sample point is calculated using the closest sample points from the original signal. This results in an N-valued vector per voltage period. Equation 8 shows the kth period vector extracted from the resampling signal S_R.

$$P_k := (\text{matrix}(S_R[k \cdot N + 1]@S_R[k \cdot N + 2]@.@.@.S_R[(k + 1) \cdot N])),$$
$$k \in \{0, \ldots, K - 1\}$$

$$P_k := \begin{pmatrix} S_R[k \cdot N + 1] \\ S_R[k \cdot N + 2] \\ \vdots \\ S_R[(k+1) \cdot N] \end{pmatrix}, \quad k \in \{0, \dots, K-1\} \tag{8}$$

Each vector P_k is appended as a row vector to the transformed voltage signal, so the FIT-PS signal in Eq. 9 results in a matrix.

$$S_{FITPS} := \left(\text{matrix}\left(P_1' @ P_2' @.@.@.@ P_k' @ P_{k+1}' @.@..\right)\right)$$

$$S_{FITPS} := \begin{pmatrix} P_1' \\ P_2' \\ \vdots \\ P_k' \\ P_{k+1}' \\ \vdots \end{pmatrix} \tag{9}$$

In Fig. 7 the transformation of the original signal S_{org} into the resampling signal form S_R is sketched.

In the next step, the same interpolation is applied to the current signal at the previously determined times. This also results in a vector of length N for the current signal,

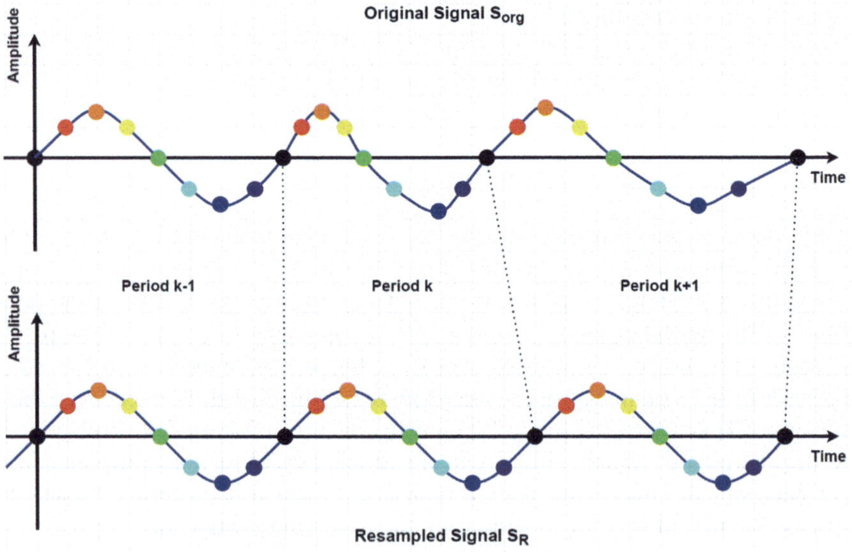

Fig. 7 Frequency-invariant resampling of the original signal

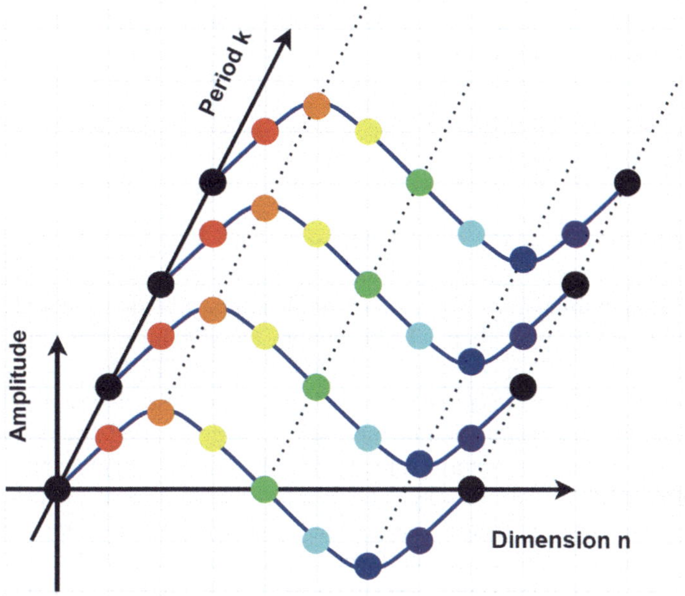

Fig. 8 Signal in FIT-PS representation

which is appended to the transformed current signal as a line vector. In Fig. 8 this transformed signal matrix can be seen as a three-dimensional FIT-PS representation.

Another form of representation is a heat map, as can be found in Held et al. (2019a), where the amplitude of the signal determines the color. The resulting, transformed signals now correspond to a fundamental frequency that corresponds exactly to the ideal mains frequency f_0. This means that relatively the same points in time within the different periods can now be compared directly. This makes it also possible, for example, to use algorithms from image processing. Likewise, phase changes in the signal are visible. Another special feature of the FIT-PS algorithm is that it can also be used for upsampling and downsampling. The resampling frequency can be determined by selecting the N parameter. This has advantages, for example, when different data sets are to be combined and compared, as shown in Held et al. (2019b).

6 Artificial Intelligence for Appliance Classification

Artificial intelligence plays one of the central roles at NILM. There are approaches in which the overall signal is evaluated, and the event detection and classification are carried out together. Other approaches assume the recognized events and only examine the relevant signal sections for their current consumers. When detecting appliances using artificial intelligence, a distinction can be made between supervised and unsupervised learning approaches.

6.1 Supervised Learning

Supervised learning requires training and test data. A special machine learning problem is solved here by training a machine learning model with existing data. There are many publications that use simple machine learning methods, but also achieve good results with them. These simple classification methods include k Nearest Neighbors (kNN), Support Vector Machine (SVM), Decision Tree, Naive Bayes, and Random Forest (Gurbuz et al. 2021; Lin and Tsai 2011; Weißhaar et al. 2018). These simple classification methods are suitable as good comparison values for more complex machine learning models. Problem-solving using neural networks is en vogue these days. Accordingly, there is many research work about it. All types of neural networks can be found, such as back propagation neural networks, recurrent neural networks (RNN), convolutional neural networks (CNN), as well as more advanced forms such as long short-term memory (LSTM) and others (Ciancetta et al. 2021; Held et al. 2019a; Le et al. 2016; Wang and Yin 2017). Techniques such as transfer learning (Devlin and Hayes 2019; D'Incecco et al. 2020; Zhou et al. 2021) are also taken into account in research, which can significantly accelerate the learning process. A problem in the context of NILM is the lack of data, which makes it difficult to train deep neural networks with good results and reduces the risk of overfitting. A possible solution is the use of data augmentation (Rafiq et al. 2021).

Both the simple machine learning algorithms and the neural networks deliver good results in the problems they examine. This is mostly because they are trained on a specific problem. The limits of the supervised learning approach come to light when it comes to problems such as newly added devices. Here the area of supervised learning must be left and looked in the direction of unsupervised learning.

6.2 Unsupervised Learning

Unsupervised learning takes a completely different perspective on device detection. In this case, it is assumed that much to all information about the electronic network and consumers is missing. Accordingly, strategies have to be applied to piecewise decompose the present overall signal and to identify consumers contained therein. In an overview of unsupervised learning in the context of NILM, Bonfigli et al. (2015) offers the idea of carrying out this device detection in two steps. In the first step, individual loads are detected. This detection is still completely detached from a specific device assignment. The second step is clustering, in which the previously identified loads are assigned to a common source. This source then ultimately corresponds to a device. The first step is pursued using various fHMM-based approaches.

The second step of actual device assignment is done here either using various forms of matrix factorization or a genetic k-means clustering method. The approaches based on fHMMs are all eventless. However, there are also event-based unsupervised learning approaches that require as little prior knowledge as possible. Kamoto et al. (2017) use a Competitive Agglomeration (CA) algorithm in their work. This clustering method starts with a too high number of possible clusters and optimizes this number. The advantage over many other clustering methods is that no a priori knowledge of the number of devices is required. The feature sequences are previously extracted from the overall signal by an event detector and then entered into the CA. In the next step, the ON and OFF clusters that belong together are identified so that simple Type I appliances can be modeled from them. From the cluster pairs found in this way, the device is then recognized in the overall signal. With the help of these detected devices, the last step is to check whether the overall performance can now be reconstructed from these individual device performances. A limitation of Kamoto et al.'s (2017) method compared to the approaches described in Bonfigli et al. (2015) is that it is generally assumed here that everything can be put together from simple devices. Each switch-on event is converted to the same switch-off event. A problem could arise with finite state machines, which allow transitions between the individual states so that they no longer run back in the same order. Likewise, Type III Various Power devices are not included in this approach. In a direct comparison, the CA-based approach of Kamoto et al. (2017) performs better overall than fHMM approaches when evaluating part of the REDD dataset, according to their investigation.

6.3 Semi-supervised and Online Learning

Compared to supervised learning approaches, unsupervised learning approaches are more generic. In supervised learning, a model is always optimized for a specific problem. This model is then complete. Subsequent addition or removal of a device will result in the model having to be retrained. Unsupervised learning approaches start from scratch and develop a model that is as suitable as possible in an existing environment. In order to be able to tackle the real problems with NILM in the future, one approach could be to develop methods that are less based on batch learning and involve more online learning. A parallel structure is conceivable here. A trained model performs the classification. Another model uses the incoming features to check whether the current clusters are still correct or whether a new cluster has to be established due to an added device. This newly defined cluster must then be integrated into the classification model via an update process, and existing clusters must be modified if necessary.

Semi-supervised and online learning approaches can already be found in various works. Egarter et al. (2015) describes an approach based on fHMM with particle filtering. In the presentation of the algorithm, there is also a description of how this algorithm can be supplemented with online learning. In Salem and Sayed-Mouchaweh (2020), a semi-supervised online learning approach based on a conditional HMM (CHMM) and the Expectation Maximization (EM) algorithm is pursued, in which an existing model is improved with continuous classifications.

7 Conclusion

This chapter gave an overview of Non-Intrusive Load Monitoring (NILM). The overall performance is measured at a central node within an electronic network. This overall signal is then broken down into its components using various algorithms, and the power is assigned to the consumers. This process is called disaggregation. The realization of NILM can be done with many methods. A standard processing chain was presented. Following this process chain, one of the first questions is how the measurement data is acquired. The measurement data are further processed, and various features can be extracted. Depending on the sampling rate of the measurement signal, low or high-frequency features must be paired with suitable event detection methods and classification methods. A fundamental distinction is made between event-based and eventless approaches. The eventless concept is usually based on factorial Hidden Markov Models. Various measurement principles were introduced which are suitable for a non-intrusive measurement. In addition to different approaches to the measurement hardware, publicly accessible data sets for NILM were presented. Various used feature extraction methods for event detection and classification have been shown. Additionally, the Frequency Invariant Transformation of Periodic Signals (FIT-PS) algorithm, developed for NILM, has been explained. Different artificial intelligence approaches for NILM were presented, divided into supervised and unsupervised learning. Finally, semi-supervised and online learning approaches with example implementations in NILM were shown.

References

Abubakar I, Khalid SN, Mustafa MW, Shareef H, Mustapha M (2015) An overview of non-intrusive load monitoring methodologies. In: 2015 IEEE conference on energy conversion (CENCON), pp 54–59. https://doi.org/10.1109/CENCON.2015.7409513

Adabi A, Manovi P, Mantey P (2016) Cost-effective instrumentation via NILM to support a residential energy management system. In: 2016 IEEE international conference on consumer electronics (ICCE), pp 107–110. https://doi.org/10.1109/ICCE.2016.7430540

Anderson K, Ocneanu A, Benitez D, Carlson D, Rowe A, Berges M (2012) BLUED: a fully labeled public dataset for event-based non-intrusive load monitoring research. In: Proceedings of the 2nd KDD workshop on data mining applications in sustainability (SustKDD), vol 7. ACM, New York, pp 1–5

Barbero JC, Hernández A, Ureña J (2020) FPGA-based architecture for identification algorithms in NILM techniques. In: 2020 IEEE international instrumentation and measurement technology conference (I2MTC), pp 1–5. https://doi.org/10.1109/I2MTC43012.2020.9128538

Beckel C, Kleiminger W, Cicchetti R, Staake T, Santini S (2014) The ECO data set and the performance of non-intrusive load monitoring algorithms. In: Proceedings of the 1st ACM conference on embedded systems for energy-efficient buildings, pp 80–89

Bergés M, Soibelman L, Matthews HS (2010) Leveraging data from environmental sensors to enhance electrical load disaggregation algorithms. In: Proceedings of the 13th international conference on computing in civil and building engineering, vol 30, Nottingham

Bonfigli R, Squartini S, Fagiani M, Piazza F (2015) Unsupervised algorithms for non-intrusive load monitoring: an up-to-date overview. In: 2015 IEEE 15th international conference on environment and electrical engineering (EEEIC), pp 1175–1180. https://doi.org/10.1109/EEEIC.2015.7165334

Bousbiat H, Klemenjak C, Leitner G, Elmenreich W (2020) Augmenting an assisted living lab with non-intrusive load monitoring. In: 2020 IEEE international instrumentation and measurement technology conference (I2MTC), pp 1–5. https://doi.org/10.1109/I2MTC43012.2020.9128406

Bucci G, Ciancetta F, Fiorucci E, Mari S, Fioravanti A (2021) State of art overview of non-intrusive load monitoring applications in smart grids. Meas Sens 18:100145. https://doi.org/10.1016/j.measen.2021.100145

Cardenas A, Agbossou K, Guzmán C (2016) Development of real-time admittance analysis system for residential load monitoring. In: 2016 IEEE 25th international symposium on industrial electronics (ISIE), pp 696–701. https://doi.org/10.1109/ISIE.2016.7744974

Ciancetta F, Bucci G, Fiorucci E, Mari S, Fioravanti A (2021) A new convolutional neural network-based system for NILM applications. IEEE Trans Instrum Meas 70:1–12. https://doi.org/10.1109/TIM.2020.3035193

Devlin M, Hayes BP (2019) Non-intrusive load monitoring using electricity smart meter data: a deep learning approach. In: 2019 IEEE power and energy society general meeting (PESGM), pp 1–5. https://doi.org/10.1109/PESGM40551.2019.8973732

D'Incecco M, Squartini S, Zhong M (2020) Transfer learning for non-intrusive load monitoring. IEEE Trans Smart Grid 11(2):1419–1429. https://doi.org/10.1109/TSG.2019.2938068

Dinesh C, Nettasinghe BW, Godaliyadda RI, Mervyn PB, Ekanayake JE, Wijayakulasooriya JV (2016) Residential appliance identification based on spectral information of low frequency smart meter measurements. IEEE Trans Smart Grid 7(6):2781–2792. https://doi.org/10.1109/TSG.2015.2484258

Egarter D, Bhuvana VP, Elmenreich W (2015) PALDi: online load disaggregation via particle filtering. IEEE Trans Instrum Meas 64(2):467–477. https://doi.org/10.1109/TIM.2014.2344373

Gao J, Giri S, Kara EC, Bergés M (2014) PLAID: a public dataset of high-resolution electrical appliance measurements for load identification research: demo abstract. In: BuildSys'14. Association for Computing Machinery, New York, NY, pp 198–199. https://doi.org/10.1145/2674061.2675032

Ghahramani Z, Jordan M (1995) Factorial hidden Markov models. In: Touretzky D, Mozer MC, Hasselmo M (eds) Advances in neural information processing systems, vol. 8. MIT Press. https://proceedings.neurips.cc/paper/1995/file/4588e674d3f0faf985047d4c3f13ed0d-Paper.pdf

Gurbuz FB, Bayindir R, Vadi S (2021) Comprehensive non-intrusive load monitoring process: device event detection, device feature extraction and device identification using KNN, random forest and decision tree. In: 2021 10th international conference on renewable energy research and application (ICRERA), pp 447–452. https://doi.org/10.1109/ICRERA52334.2021.9598679

Hart GW (1989) Residential energy monitoring and computerized surveillance via utility power flows. IEEE Technol Soc Mag 8(2):12–16. https://doi.org/10.1109/44.31557

Hart GW (1992) Nonintrusive appliance load monitoring. Proc IEEE 80(12):1870–1891. https://doi.org/10.1109/5.192069

Held P, Laasch F, Abdeslam DO, Benyoucef D (2016) Frequency invariant transformation of periodic signals (FIT-PS) for signal representation in NILM. In: IECON 2016—42nd annual conference of the IEEE industrial electronics society, pp 5149–5154. https://doi.org/10.1109/IECON.2016.7793617

Held P, Mauch S, Saleh A, Benyoucef D, Abdeslam DO (2018a) HELD1: home equipment laboratory dataset for non-intrusive load monitoring. In: SIGNAL 2018 editors, p 23

Held P, Weißhaar D, Mauch S, Abdeslam DO, Benyoucef D (2018b) Parameter optimized event detection for NILM using frequency invariant transformation of periodic signals (FIT-PS). In: 2018 IEEE 23rd international conference on emerging technologies and factory automation (ETFA), vol 1, pp 832–837. https://doi.org/10.1109/ETFA.2018.8502522

Held P, Mauch S, Saleh A, Abdeslam DO, Benyoucef D (2019a) Frequency invariant transformation of periodic signals (FIT-PS) for classification in NILM. IEEE Trans Smart Grid 10(5):5556–5563. https://doi.org/10.1109/TSG.2018.2886849

Held P, Weißhaar D, Abdeslam DO, Benyoucef D (2019b) Generation of new simulation scenarios for NILM based on real data sets using high-resolution current waveforms. In: IECON 2019—45th annual conference of the IEEE industrial electronics society, vol 1, pp 5319–5324. https://doi.org/10.1109/IECON.2019.8926895.24

Held P, Weißhaar D, Abdeslam DO, Benyoucef D (2020) Investigation of Rogowski current sensors for appliances classification in NILM. In: 2020 IEEE 3rd international conference and workshop in Óbuda on electrical and power engineering (CANDO-EPE), pp 000027–000032. https://doi.org/10.1109/CANDO-EPE51100.2020.9337802

Hoyo-Montaño JA, Pereyda-Pierre CA, Tarín-Fontes JM, Leon-Ortega JN (2016) Overview of non-intrusive load monitoring: a way to energy wise consumption. In: 2016 13th international conference on power electronics (CIEP), pp 221–226. https://doi.org/10.1109/CIEP.2016.7530760

Kahl M, Ul Haq A, Kriechbaumer T, Jacobsen H-A (2016) WHITED—a worldwide household and industry transient energy data set. In: 3rd international workshop on non-intrusive load monitoring, pp 1–4

Kamoto KM, Liu Q, Liu X (2017) Unsupervised energy disaggregation of home appliances. In: Sun X, Chao H-C, You X, Bertino E (eds) Cloud computing and security. Springer International Publishing, Cham, pp 398–409

Kelly J, Knottenbelt W (2015) The UK-DALE dataset, domestic appliance-level electricity demand and whole-house demand from five UK homes. Sci Data 2(1):1–14

Klein P, Merckle J, Benyoucef D, Bier T (2013) Test bench and quality measures for non-intrusive load monitoring algorithms. In: IECON 2013—39th annual conference of the IEEE industrial electronics society, pp 5006–5011. https://doi.org/10.1109/IECON.2013.6699946

Kolter JZ, Johnson MJ (2011) REDD: a public data set for energy disaggregation research. In: Workshop on data mining applications in sustainability (SIGKDD), vol 25, San Diego, CA. Citeseer, pp 59–62

Le T-T-H, Kim J, Kim H (2016) Classification performance using gated recurrent unit recurrent neural network on energy disaggregation. In: 2016 international conference on machine learning and cybernetics (ICMLC), vol 1, pp 105–110. https://doi.org/10.1109/ICMLC.2016.7860885

Leferink F, Keyer C, Melentjev A (2016) Static energy meter errors caused by conducted electromagnetic interference. IEEE Electromagn Compat Mag 5(4):49–55. https://doi.org/10.1109/MEMC.2016.7866234

Liang J, Ng SKK, Kendall G, Cheng JWM (2010) Load signature study—part I: basic concept, structure, and methodology. IEEE Trans Power Delivery 25(2):551–560. https://doi.org/10.1109/TPWRD.2009.2033799

Liang M, Meng Y, Lu N, Lubkeman D, Kling A (2019) HVAC load disaggregation using low-resolution smart meter data. In: 2019 IEEE power energy society innovative smart grid technologies conference (ISGT), pp 1–5. https://doi.org/10.1109/ISGT.2019.8791578

Lin Y-H, Tsai M-S (2011) Applications of hierarchical support vector machines for identifying load operation in nonintrusive load monitoring systems. In: 2011 9th world congress on intelligent control and automation, pp 688–693. https://doi.org/10.1109/WCICA.2011.5970603

Lu M, Li Z (2020) A hybrid event detection approach for non-intrusive load monitoring. IEEE Trans Smart Grid 11(1):528–540. https://doi.org/10.1109/TSG.2019.2924862

Luca G, Benedetta M, Nardecchia F, Bisegna F, Chiara G (2015) Home smart grid device for energy saves and failure monitoring. In: 2015 IEEE 15th international conference on environment and electrical engineering (EEEIC), pp 671–676. https://doi.org/10.1109/EEEIC.2015.7165245

Martins JF, Lopes R, Lima C, Romero-Cadaval E, Vinnikov D (2012) A novel nonintrusive load monitoring system based on the s-transform. In: 2012 13th international conference on optimization of electrical and electronic equipment (OPTIM), pp 973–978. https://doi.org/10.1109/OPTIM.2012.6231777

Morán A, Alonso S, Pérez D, Prada MA, Fuertes JJ, Domínguez M (2020) Feature extraction from building submetering networks using deep learning. Sensors 20(13). https://doi.org/10.3390/s20133665

Pereira L, Nunes N (2018) Performance evaluation in non-intrusive load monitoring: datasets, metrics, and tools—a review. Wiley Interdiscip Rev Data Min Knowl Discov 8(6):e1265

Picon T, Meziane MN, Ravier P, Lamarque G, Novello C, Le Bunetel J-C, Raingeaud Y (2016) COOLL: controlled on/off loads library, a public dataset of high-sampled electrical signals for appliance identification. CoRR. http://arxiv.org/abs/1611.05803

Rafiq H, Shi X, Zhang H, Li H, Ochani MK, Shah AA (2021) Generalizability improvement of deep learning-based non-intrusive load monitoring system using data augmentation. IEEE Trans Smart Grid 12(4):3265–3277. https://doi.org/10.1109/TSG.2021.3082622

Raiker GA, Reddy SB, Umanand L, Yadav A, Shaikh MM (2018) Approach to non-intrusive load monitoring using factorial hidden Markov model. In: 2018 IEEE 13th international conference on industrial and information systems (ICIIS), pp 381–386. https://doi.org/10.1109/ICIINFS.2018.8721436

Ruano A, Hernandez A, Ureña J, Ruano M, Garcia J (2019) NILM techniques for intelligent home energy management and ambient assisted living: a review. Energies 12(11). https://doi.org/10.3390/en12112203

Salem H, Sayed-Mouchaweh M (2020) A semi-supervised and online learning approach for non-intrusive load monitoring. In: Brefeld U, Fromont E, Hotho A, Knobbe A, Maathuis M, Robardet C (eds) Machine learning and knowledge discovery in databases. Springer International Publishing, Cham, 585–601

Shiddieqy HA, Hariadi FI, Adijarto W (2021) Plug-load classification based on CNN from V-I trajectory image using STM32. In: 2021 international symposium on electronics and smart devices (ISESD), pp 1–5. https://doi.org/10.1109/ISESD53023.2021.9501919

Trung KN, Zammit O, Dekneuvel E, Nicolle B, Van Nguyen C, Jacquemod G (2012) An innovative non-intrusive load monitoring system for commercial and industrial application. In: The 2012 international conference on advanced technologies for communications, pp 23–27. https://doi.org/10.1109/ATC.2012.6404221

Wang TY, Yin B (2017) A new method for the nonintrusive load monitoring based on BP neural network. In: 2017 2nd international conference on multimedia and image processing (ICMIP), pp 93–97. https://doi.org/10.1109/ICMIP.2017.55

Weißhaar D, Held P, Mauch S, Benyoucef D (2018) Device classification for NILM using FIT-PS compared with standard signal forms. In: 2018 international IEEE conference and workshop in Óbuda on electrical and power engineering (CANDO-EPE), pp 1–6. https://doi.org/10.1109/CANDO-EPE.2018.8601150

Weißhaar D, Held P, Abdeslam DO, Benyoucef D (2020) Expansion and superposition of switching cycles to generate simulation datasets for NILM. In: IECON 2020 the 46th annual conference of the IEEE industrial electronics society. IEEE, pp 5163–5169

Wild B, Barsim KS, Yang B (2015) A new unsupervised event detector for non-intrusive load monitoring. In: 2015 IEEE global conference on signal and information processing (GlobalSIP), pp 73–77. https://doi.org/10.1109/GlobalSIP.2015.7418159

Yaemprayoon S, Boonplian V, Srinonchat J (2016) Developing an innovation smart meter based on CS5490. In: 2016 13th international conference on electrical engineering/electronics, computer, telecommunications and information technology (ECTI-CON), pp 1–4. https://doi.org/10.1109/ECTICon.2016.7561400

Yang D, Gao X, Kong L, Pang Y, Zhou B (2020) An event-driven convolutional neural architecture for non-intrusive load monitoring of residential appliance. IEEE Trans Consum Electron 66(2):173–182. https://doi.org/10.1109/TCE.2020.2977964

Yang F, Liu B, Luan W, Zhao B, Liu Z, Xiao X, Zhang R (2021) FHMM based industrial load disaggregation. In: 2021 6th Asia conference on power and electrical engineering (ACPEE), pp 330–334. https://doi.org/10.1109/ACPEE51499.2021.9436945

Zeifman M, Roth K (2011) Nonintrusive appliance load monitoring: review and outlook. IEEE Trans Consum Electron 57(1):76–84. https://doi.org/10.1109/TCE.2011.5735484

Zhang L, Zhu L (2019) Basic summary of non-intrusive load monitoring. In: 2019 IEEE 10th international conference on software engineering and service science (ICSESS), pp 1–5. https://doi.org/10.1109/ICSESS47205.2019.9040726

Zhou Z, Xiang Y, Xu H, Yi Z, Shi D, Wang Z (2021) A novel transfer learning-based intelligent nonintrusive load-monitoring with limited measurements. IEEE Trans Instrum Meas 70:1–8. https://doi.org/10.1109/TIM.2020.3011335

Zhuang M, Shahidehpour M, Li Z (2018) An overview of non-intrusive load monitoring: approaches, business applications, and challenges. In: 2018 international conference on power system technology (POWERCON), pp 4291–4299. https://doi.org/10.1109/POWERCON.2018.8601534

Zoha A, Gluhak A, Imran MA, Rajasegarar S (2012) Non-intrusive load monitoring approaches for disaggregated energy sensing: a survey. Sensors 12(12):16838–16866. https://doi.org/10.3390/s121216838

Helping Consumers to Reduce Their Energy Consumption and Greenhouse Gases Emissions: What Tool to Develop?

Lorris Tabbone, Nadège Blond, Clémentine Ciani, Jona Prifti, Paul Salze, and Sandrine Glatron

1 Introduction

Facing climate change, caused by human activities largely through the combustion of fossil fuels which releases large quantities of greenhouse gases into the atmosphere, we need to transform our energy systems. Energy saving is an essential condition for this transformation. Indeed, intermittent renewable energies cannot replace fossil fuels to meet the current energy demand yet without developing expensive and polluting storage infrastructures. These objectives are included in the agenda of the European Union. This agenda is declined in France with the law of 17 August 2015 on the energy transition for green growth (LTECV Loi relative à la Transition Energétique pour la Croissance Verte). Participation of all territorial stakeholders is expected to foster less energy-intensive practices and, in general, more sustainable behaviors to reduce impacts on the biosphere over the long term (Lutzenhiser and Gossard 2000; Brisepierre 2011; Brounen et al. 2012).

Numerous tools are proposed to foster this energy and ecological transition: investments in renewable energy production, clean transport, low-energy building conception and renovation, sustainable agriculture, tax incentives, new norms and advice

L. Tabbone · N. Blond (✉) · C. Ciani · J. Prifti · P. Salze
Laboratoire Image Ville Environnement, Université de Strasbourg, CNRS, UMR 7362, Strasbourg, France
e-mail: nadege.blond@live-cnrs.unistra.fr

J. Prifti
e-mail: joprifti@hotmail.com

L. Tabbone · C. Ciani · J. Prifti · S. Glatron
Laboratoire Interdisciplinaire en Études Culturelles, Université de Strasbourg, CNRS, UMR 7069, Strasbourg, France

L. Tabbone · N. Blond · C. Ciani · J. Prifti · P. Salze · S. Glatron
LTSER France, Zone Atelier Environnementale Urbaine, 3 Rue de l'Argonne, 67000 Strasbourg, France

D. Ould Abdeslam (ed.), *Smart Meters*, Lecture Notes in Energy 97,
https://doi.org/10.1007/978-3-031-27556-2_4

on energy efficiency in all sectors, and communication campaigns to promote eco-friendly behavior, etc. Among them, smart meters and associated applications are spread out to help households to better control their energy consumption. Despite their importance for energy management and saving, smart meters aren't well accepted. The reasons are to be found in many obstacles such as knowledge, values, norms, or beliefs (Schweitzer and Simon 2021). Moreover, the expected change in users' consumption behavior is not guaranteed (Zélem 2010): Although consumers seem to be aware of the necessity and the urgency to reinforce the energy and ecological transition, they do not engage in real transformations of their consumption mechanisms, which are too strongly linked to their habits. Finding issues to faster the transformations need more focused attention and research.

Several applications linked to smart meters are currently available and may guide the consumers in reducing their energy consumption. They take many forms but are mostly underused. We hypothesize that these applications are not enough identified as important instruments for the energy and ecological transition. They are rather seen as spyware-like tools controlled by political and economic processes. The lack of transparency of the tools and their providers is probably an important issue.

The objective of this chapter is first to analyze such applications made available to consumers. Since the latter's degree of awareness of energy and ecological issues may influence their decision to use or not to use these applications, a more in-depth analysis of consumers' representations of the energy and ecological issues, as well as the energy planning/policies/tools is proposed based on outcomes of both qualitative and quantitative surveys. These representations are confronted with overall energy consumption behaviors.

We state that citizens must no longer be considered simple consumers whose energy consumption only responds to technical and economic logic. Their energy needs are controlled by their environment and their lifestyle, which influence is important to be highlighted to make citizens aware of their potential to reduce their footprints. Individual lifestyles and practices are essential factors to consider in the energy-saving policies and associated tools. Developing the ecological awareness of individuals is an essential step to motivate their energy-saving for environmental reasons and not only for financial ones (cost saving can have direct and indirect rebound effects). In this way, the chapter also discusses the structural implementation of the SMI web interface, designed to help users in their energy savings along with several steps of ecological awareness. The tool aims at fostering the overall "ecological reflexivity" of energy consumers, defined as the degree of an individual's awareness of his or her consumption practices and their impact on the environment.

Section 2 presents the analytical framework for our study. Section 3 discusses the results of an analysis of the energy management tools, and the main findings of two qualitative and quantitative surveys on the representation of energy management. Section 4 presents the structuration of an SMI web interface. The section concludes with the work carried out.

2 Practices and Control of Energy Consumption

The individual practices, inducing energy consumption, are influenced by several factors which are detailed according to socio-demographic, geographical, and technical dimensions in the scientific literature depending on the approach, discipline, and issue. This section presents the cross-cutting 'lifestyle' approach that describes such individual practices bringing together all the propositions. This lifestyle approach is guiding our analyzes as well as the SMI web interface development oriented to make the households aware of the role of their practices in their energy consumption patterns and contribute to favor their energy savings.

2.1 Analysis of Energy Consumption by Lifestyle

The study of energy consumption through "lifestyle" consists in considering all the individual and social factors that influence citizens' practices. Herpin and Verger (2008) define the lifestyle 'as a matrix that creates needs': individuals take positions and decisions concerning themselves and/or their family (professional choices, residential location, choice of activities, etc.) that condition their practices and thus their energy consumption.

Most of the studies based on the lifestyle approach link the lifestyles and energy consumption through the household composition, the description of the owned equipment, their use frequencies, their time of use, and the individual beliefs (Lutzenhiser 1993). Some authors highlight that income is a good indicator of energy consumption: Roy (2007) shows that the higher the consumer's income, the more he consumes, even though the consumer remains sensitive to ecological discourse and in favor of energy savings. Consumption is then mainly identified as a social indicator. Druckman and Jackson (2008), Brisepierre (2011) and Sahakian (2011) show that the household energy consumption also depends on the type and location of the inhabitants in addition to social and cultural factors. According to Gram-Hanssen and Bech-Danielsen (2004), lifestyles influence the choice of dwelling. Sanquist et al. (2012) explain that households are positioned in a socio-spatial living space: This position changes when they marry and have children when they move to another location; these life modifications induce changes in the intensity and frequency of their energy consumption.

A sample of more than 300,000 Dutch households was studied by Brounen et al. (2012). They show that gas consumption is mainly controlled by the structural characteristics of the dwelling (age, type of building), while electricity consumption varies more according to the composition of the household (especially income and family structure). The EDF R&D (Electricité De France Research and Development) team states that technical and structural factors are responsible for 2/3 of the explained variance of the heating energy consumption while 1/3 of the variance is explained by household socio-demographic variables (Cayla et al. 2010). Tabbone (2017) studied the life cycle of households during the time, their lifestyle (social dispositions and practices), and the links with their living environment (characteristics of the dwelling and its surroundings). He concluded that lifestyle guides the individuals towards the desired living environment and in return, this residential choice, associated with the technical offer of the chosen territory, influences their lifestyle. Indeed, individuals have activities according to the possibilities territory offers.

The analysis of practices inducing energy consumption is usually complex as it depends on many individual and social factors for which there is a lack of data. Moreover, diverse fields of expertise (sociology, geography, engineering, etc.) are necessary to study the same time the legislation, the technical offer of the territory, the housing, the behavior, the energy consumption, etc. Ideally, it would be useful to provide tools that would provide the consumers or the stakeholders a direct comparison of the energy consumption with explanatory factors, including information on the lifestyles and environments, and a proposition of adapted solutions for energy savings.

2.2 *Framework for Analyzing Practices and Representations of Energy Demand Management (EDM)*

To analyze the practices of the individuals, and the related energy consumption, explanatory factors are derived from previous studies, classified according to two dimensions, the social dispositions, and the living environment.

- **The social dispositions** of an individual refer to his or her values, habits, attitudes—opinions/biases/beliefs. They are strongly dependent on the social capital (social environment of the individual, access to networks, etc.), the economic capital (incomes), and the cultural capital (knowledge, culture, etc.). Social dispositions are built gradually over time and are strongly influenced by the individual's network (family, friends, etc.). They are internalized by the individual throughout his or her life. Although they are not fixed, they remain difficult to change.
- **The living environment** describes the individual's environment, including his or her dwelling (house/apartment, rural/urban, old/new) and all the related equipment, the socio-technical offer of the inhabited territory (housing offer, heating system, competence of professionals, costs of what is offered), public policies (disincentives, obligations/prohibitions, etc.), the local climate.

The social dispositions and the living environment of an individual co-evolve during his or her life cycle (change his or her professional and family context: getting married, having children, separating, or divorcing, etc.) modifying his or her practices, and the related energy consumption (Fig. 1).

Energy Demand Management (EDM) refers to a set of measures aimed at saving energy. These measures can be implemented by various actors to limit the energy demand, i.e. all energy consumption. Consumers themselves are targeted by these

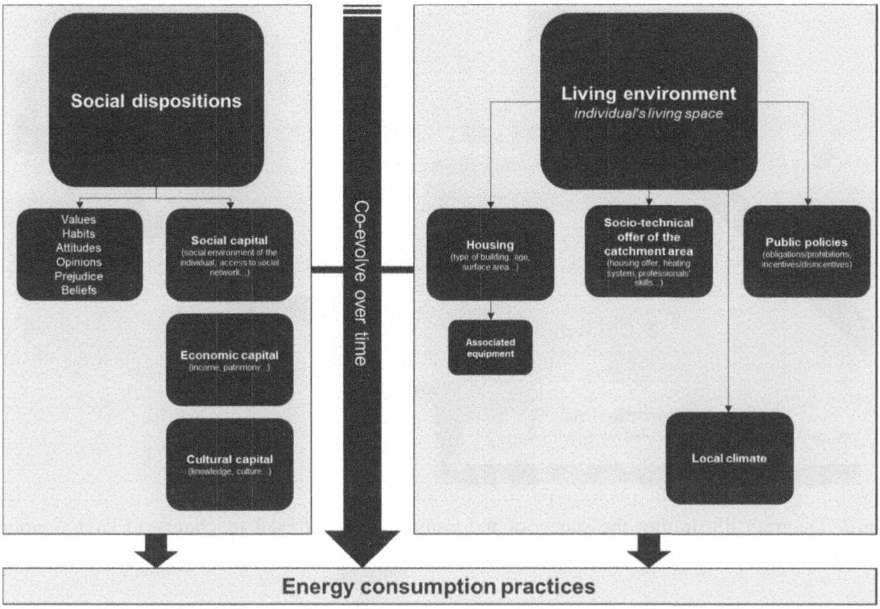

Fig. 1 Representative diagram of the link between social dispositions and the living environment and its effect on energy consumption practices. *Source* Authors

measures to limit their energy demand. EDM measures aim at modifying the energy consumption trajectory and the practices of individuals towards greater energy savings. Zélem (2010) shows that today's regulatory and technical measures are still limited and should better consider the uses, practices, and representations of these different actors.

2.3 *Energy Demand Management (EDM) Measures*

Energy demand management (EDM) includes communication campaigns, challenges, financial incentives such as off-peak/peak hours, and applications (websites, mobile phone applications, etc.) enabling consumers to analyze their energy consumption. Smart meters and their associated web applications are EDM tools deployed in the territories to organize automatic monitoring of household electricity consumption and more efficient energy planning strategies. Schweitzer and Simon (2021, task 2 of the Interreg SMI project) show that such tools are not well accepted because of many obstacles related to the social dispositions of individuals: knowledge, attitudes, norms, values, and beliefs of consumers. They explain that fostering the contribution of the users in the development and the implementation of the tool

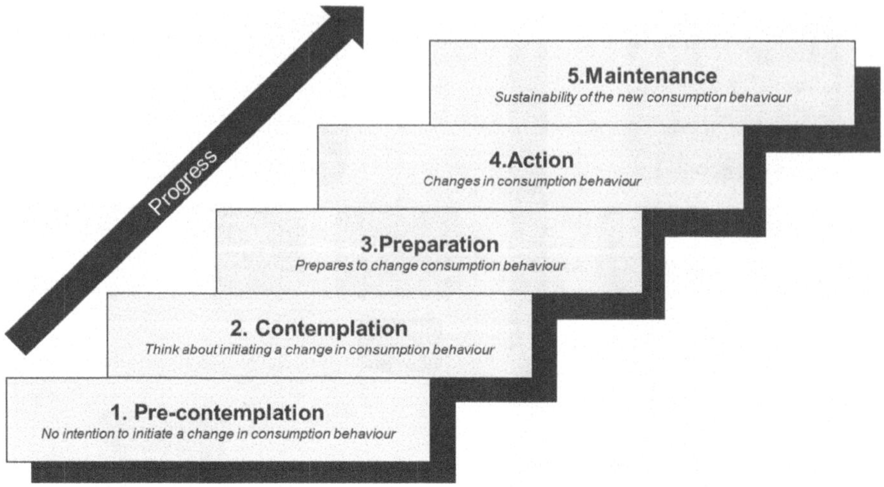

Fig. 2 Diagram illustrating the stages of the transtheoretical model of change (TTM). *Source* Authors via Prochaska and Velicer (1997)

would favor the use of the smart meter, especially if positive symbolic representations are associated with it. This would help the user to understand its functioning and would limit the "privacy paradox", which refers to the existing "gap between stated privacy preferences and actual usage behavior given the identified benefits".

AlSkaif et al. (2018) explain that consumer acceptance and use of such tools follow several steps. They proposed the 'transtheoretical model of change' (TTM) (Prochaska and Velicer 1997) that they adapted to the study of energy consumption behavior (Fig. 2). Five steps describe the intellectual processes occurring when using an EDM tool. The pathway is full of obstacles that the tool may help to overcome to orient the user on an energy-saving trajectory.

The first step is a pre-contemplation phase describing a consumer with no intention of changing his or her consumption behavior. The tool must provide information to the user for him or her to identify the mechanism of his or her actions and their consequences on his or her energy consumption, whatever are the results. The second step considers a behavior change. The EDM tool must display the data collected in a fully transparent way to the consumer, protecting his or her privacy (transparency and privacy protection are two essential steps in the individual's acceptance of the tool).

The third step consists of preparing the action. The individuals must use their knowledge and experiment with the EDM tool. The tool must be fun, easy to access, and must allow the consumer to quickly observe the results of some actions. The next step is to implement a new behavior, while the final step is to maintain this new behavior. It is important then to motivate the consumers and prevent possible backtracking by providing evidence of the benefits of their efforts. Beck et al. (2019) list ten mechanisms that are thought to facilitate user motivation: point system, ranking, badge, level, presence of a story, clear goals, feedback, reward, progress, and challenge. Cost indicators are discussed by various authors (Ehrhardt-Martinez et al. 2010; Kendel 2015; Zélem 2018) who explain that the financial gains are usually quickly absorbed by greater use of electrical appliances (this is called the direct rebound effect), or the acquisition of new consumer goods (indirect rebound effect).

3 Analysis of the Tools and the Representations of the EDM

We hypothesize that the EDM tools currently available are not sufficiently used also because consumers are not enough identifying those tools as important for the energy and ecological transition. Personal dispositions are certainly key factors of explanation. Energy and ecological issues are not priorities for everyone and are not yet sufficiently integrated into the individual daily agenda. They are not enough explained by the energy suppliers through the EDM tools while adopting habits for energy saving is certainly a long process that must be supported by an awareness of the underlying issues.

Task 3.5 of the SMI project has several objectives:

- Analyze the web applications/tools provided to individuals to help them to reduce their energy consumption;
- Characterize the individuals' representation of the energy management and the associated tools;
- Propose a web interface to the SMI, developed in Task 3, that can foster the degree of an individual's awareness of his or her consumption practices and the impact of the latter on the environment (ecological reflexivity).

3.1 Analysis of EDM Tools

Applications were extracted from the Google Play platform using several keywords relating to energy-saving and planning. We note, as well as Beck et al. (2019) for other platforms, that no specific category was created for EDM tools. Most applications are proposed by electricity suppliers through a contract that allows the user to have access to an account. Whether issued from the private market sphere (such as Ecojoko) or by public providers (EcoWatt, Enedis, EDF and Moi, etc.), addressed to an "individual"—a single household/user, or a "collective"—open to all, applications were analyzed. In the text presenting each application, we detail the provider, the feedback (individual or collective), and the presence or absence of references to the energy underlying issues (climate change, energy planning, technical issues related to the occurrence of "blackouts", etc.), the indicators used to describe the household (kWh, Watt, Euro, grams of CO_2, etc.), the type of actions proposed to save energy, the presence or absence of games.

The analysis shows that arguments in favor of ecology and carbon footprint are mostly absent from individual applications or mentioned as secondary information. On the other hand, the collective applications focus on collective contribution: "Let's act together for the planet", "Let's act against climate change", and "Change our habits to protect the planet". A "community" section lists various objectives related to climate risk ("Reduce CO_2") and technical risks ("Reduce nuclear waste", "Reduce peak demand"). Individual action is transposed into global objectives, with no direct benefit to the user.

Individual applications generally show cost indicators (in euros) and to a lesser extent the energy consumption in kWh. Some applications even promote the change of equipment and offer discounts to buy them. Collective devices present direct information on energy consumption (in watts, "mini watts"), or other indicators such as the quantity of uranium consumed (in micrograms), or the quantity of CO_2 emitted (in grams) and its equivalence in terms of activity, such as the number of kilometers traveled by a vehicle.

All the applications claim the non-degradation of the lifestyles ("while preserving your comfort", "without impacting your comfort") or refer to an environmental impact reduction "without effort". Such energy transition representation was supported by Mr. Maroš Šefčovič, Vice-President in charge of the Energy Union portfolio, who stated in 2018: "*We cannot safely live on a planet with a climate that is out of control. But that does not mean that to reduce emissions, we should sacrifice the livelihoods of Europeans.*" On the contrary, Carbone 4 explains that "*the impact of individual actions is far from negligible*" and the individual efforts "*cannot by themselves enable us to achieve the − 80% reduction in the personal carbon footprint compatible with the Paris Agreement*" without a great loss of comfort (doing without meat, cars, planes, etc.).

Like Beck et al. (2019), we noted that only a few applications, issued from the public sphere, use games to encourage consumers to reduce their energy consumption. Nevertheless, the role of games is still questionable since games may reinforce the idea that the energy transition can be easy.

Other questions arise from this work regarding the understanding of non-economic indicators (is everyone able to understand what do a *Wh*, a *gram* of uranium, or CO_2 mean?) and the ambiguous position of energy suppliers who are expected to contribute to the reduction of the energy consumptions while their profits depend on them.

3.2 What Social Representations of Energy Management?

Since the analysis of the relationship of individuals with the energy management, in general, may help to investigate the reasons why the EDM tools are not used, qualitative and quantity surveys were performed. The surveys aimed to detail the individuals' representations of climate change and energy issues, as well as of the energy management, and related tools. The main questions are: How do users associate the EDM tools with the energy and ecological transition? Can we distinguish several types of representations? Are political conceptions influencing the use of the EDM tools? Answers were expected to be useful elements to develop the web interface of the SMI.

Qualitative Survey

Thirteen qualitative interviews were first conducted in May and June 2020. This period corresponds to the first wave of the COVID-19 crisis during which the French population was locked in. The interviews were therefore conducted remotely by telephone. The impact of such interviewing conditions cannot be discussed at this stage. Table 1 lists the respondents, their gender, age, occupational category (OC), whether they live in a house or a flat, and whether they have a smart meter or not.

The questions aimed at detailing the interviewee's lifestyle and his or her relationship with the EDM tools and policies concerning the climate and energy issues.

Table 1 Characteristics of the thirteen individuals interviewed for the qualitative survey

Respondent no.	Gender	Age group	OC	Housing	Linky (SM)
1	M	50–60	Executives and higher intellectual professions	Flat	No
2	F	60+	Retired	Flat	Yes
3	M	30–40	Executives and higher intellectual professions	Flat	Yes
4	F	40–50	Intermediate occupations	Flat	Yes
5	M	20–30	Unemployed	Flat	Yes
6	F	40–50	Intermediate occupations	Flat	No
7	F	20–40	Tradesperson, shop, or business owner	House	No
8	M	40–50	Executives and higher intellectual professions	House	No
9	F	40–50	Executives and higher intellectual professions	House	No
10	M	50–60	Intermediate occupations	House	Yes
11	M	50–60	Executives and higher intellectual professions	House	Yes
12	F	40–50	Intermediate occupations	House	No
13	M	60+	Retired	Flat	Yes

To be able to describe the overall state of the ecological reflexivity of the interviewees, several questions were asked concerning their practices in activities inducing a product or energy consumption (shopping, transportation, and use of any kind of energy at home). This enabled us to know whether or not the individual makes an effort to reduce his or her carbon footprint, and if any, in which sector of activity and with which strategy (e.g. take into off-peak hours, buy energy-efficient devices, contract with green electricity provider, etc.). The EDM tools were discussed in terms of use, individual and societal impacts, or representations. The interest in the new offers of the energy market, which become more flexible in terms of the supply and distribution of the renewable energies, was evaluated. Few questions concerned the building characteristics (location, age, surface area, etc.), and the energy housing costs of the households to evaluate if the costs are "*constrained expenditure*" (Maresca et al. 2009). Do individuals make energy-saving efforts with regards to their costs? or their benefits for the environment? Questions were asked to understand the relationship of interviewees with the risks. Have users integrated the precautionary principle or the "foresight norm" (Comby and Grossetête 2012), which requires awareness of the repercussions of our actions environment? Since the underlying moral injunction is influenced by the relationship of individuals with governing bodies, we asked questions concerning evidence in policies and voting practices. Confidence in the future was also discussed.

Several types of behaviors were observed among the interviewees while no real categories could be distinguished. Some interviewees expressed doubts regarding the efforts to develop when others do not: "*We are a bit helpless. Well, I feel a bit helpless because I think that if we are the only ones doing something, nothing will change*" (Léo). The respondents feel disarmed in the face of climate change and realize that only a minority of people are aware of the ecological issues. Such interviewees showed different levels of ecological reflexivity and contrasting opinions on energy saving.

We observe that most of the behaviors are not guided by environmental concerns but much more by financial ones. Economic links to social conditions are discussed: "*Even for me, honestly, it's marginal, the optimization of electricity consumption isn't necessarily driven solely by ecological concerns, there's also the financial aspect…*" (Léo) "*I'm trying to adapt, especially since I'm on my own now, so there are no longer two incomes… I indeed try to be prudent*" (Anaïs). The interest in the energy topic and the contribution to energy saving is often linked with the representation of the energy cost weight in the global household budget. Sometimes these costs are considered too low to justify energy saving and behavior change. One example concerns the use of the off-peak system "*No, no, no, the off-peak rates, according to the literature I had read here and there, correspond to a few cents of savings, so no, we're not into that*" (Victor); "*but then, you shouldn't be a slave of the system either… I know some people who do laundry at night, to earn a few cents, but you shouldn't – well… come on! (laughs)*" (Grégoire).

Some users express distrust in politicians and their way to address climate change, while they have little knowledge of the energy efficiency and energy transition objectives: "*It's all very well to have agendas, but with all the lobbyists, once again in politics, all the agendas are either distorted or modified and, in the end, it's only 50% of the initial agenda that comes to fruition, which makes no sense*" (Grégoire); "*Politically, we're going to say 'yes, but jobs depend on it!' We have jobs, we support jobs through consumption, and consumption, whatever we say, has a carbon footprint. [...] Overall, the political actions are not strong enough to support all this*" (Elsa); "*Overall, there is a will but it's still a bit lukewarm, even at European level I don't have the impression that they have... well I don't feel that they have the desire to go into that*" (Léo).

Many respondents do not see how the EDM tools will help the energy transition. While understanding the efforts that they must develop to reduce their carbon footprint (mainly concerning transport and food), people with higher cultural capital do not perceive the solutions provided to reach energy efficiency as sufficient. The technological solutions are found useless to support the energy transition. Green technologies and energies are often pointed out: "*As soon as there is a solution that seems good, we immediately realize that there are important counterparts, in particular making cars run on electricity*" (Victor); "*And there is no green energy and so on, I mean, there are things that make you wonder, today solar panels are made from rare metals and you have to go and find them in Chile, China and so on, and that the recycling of these things is not very well organized and that solar panels are not managed today, we hear about wind turbines but it's the same thing, so today we build wind turbines, and nobody asks what we're going to do with them afterward*" (Elsa); "*Because between running a 10,000 euro car on diesel and running a 30,000 euro car on electric power, you have to have the money in the first place, and then the batteries of an electric car also pollute, so it would just be a problem that would be displaced*" (Christian). "*I don't see the point of wind power now from an energy point of view, it has been proposed to me to put some on my house, and on my neighbor's house*" (Vincent); "*oil, it's not ready to stop with shale gas, and after, etc., etc. we'll find something else, look at all the crops, which are also biofuel, from plants... who's going to settle again, so we'll deforest to make biofuel, then we'll put pesticides so that it grows better*" (Karen).

Distrust of politicians is notably observed according to the sociological profiles of the interviewees. This result is coherent with the national observations of the CEVIPOF polling institute which qualifies the 2009–2019 decade as a '*black decade for political confidence in France*'. This lack of trust in politicians and also in companies disfavor changes in energy suppliers and contracts despite the gradual liberalization of the electricity market. Far from making their choice according to a cost–benefit trade-off, individuals prefer to follow the advice of trusty third parties in their close network relations: "*so I do it and if one day in a debate with one of them [anti-nuclear activists], well at least I'll be blameless and I'll have something to talk about... that's it*".

The comparisons of energy consumption between similar households offered by a few EDM tools appear as a limitation to the energy savings, the interviewees felt to be within the norm. The interviewees also raise concerns about the security of smart meters, their applications, and the protection of personal privacy: "*There are all these debates today about global warming and what we need to do, and I find that these systems [EDM tools] are very opaque, in the end, we all sign "accepted conditions" agreements, without reading them because they are three pages long and boring, but I'm still very dubious about what they are going to do with our data...*" (Elsa).

The interviews finally revealed that smart meters and their applications help to reinforce existing energy-saving behaviors. These tools are usually used only by one member of the household, who finds satisfaction in a better understanding of energy consumption (understanding peaks, awareness of energy costs, and their lifestyles compared to others). Some users expect more information to understand how they can reduce their consumption and increase their energy production (when existing) while less active users see the energy monitoring as an additional mental burden.

Quantitative Survey

To complete the previous analysis, a quantitative survey focused on studying the representation of energy management and smart meters. The sample consisted of 164 people (90 were women and 74 were men) of whom 104 declared themselves to be executives and higher intellectual professions, 20 employees issued from administrations and services activities, 12 students, 12 with intermediate employment, 9 tradespersons, business or shop owners, 1 farmer, and the rest with non-response. The respondents answered questions designed to capture their feelings about the environment and energy issues, their belief in climate change, confidence in technology, and their representation of the EDM and associated tools. Responses to the questions are classified following a Likert scale (sometimes called a satisfaction scale) ranging from "strongly disagree" to "strongly agree". The results are presented as the number of responses as proposed by Boone and Boone (2012).

Our results show that climate change is well identified as an established process assuming that the topic is not controversial among the respondents (Fig. 3).

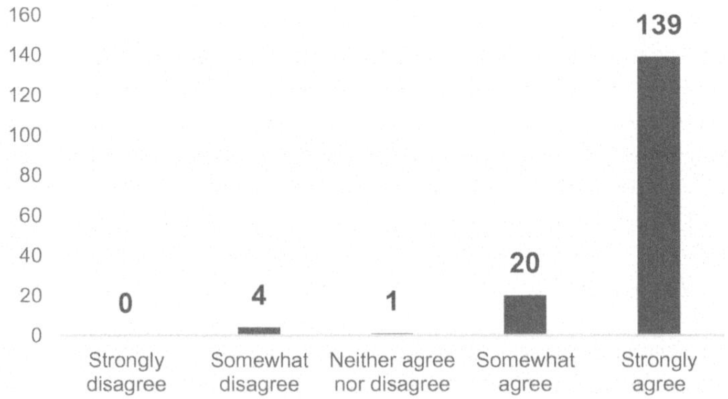

Fig. 3 Frequencies of responses to question 1 "**Climate change is a well-fact**" in the quantitative questionnaire on representations of EDM and smart meters

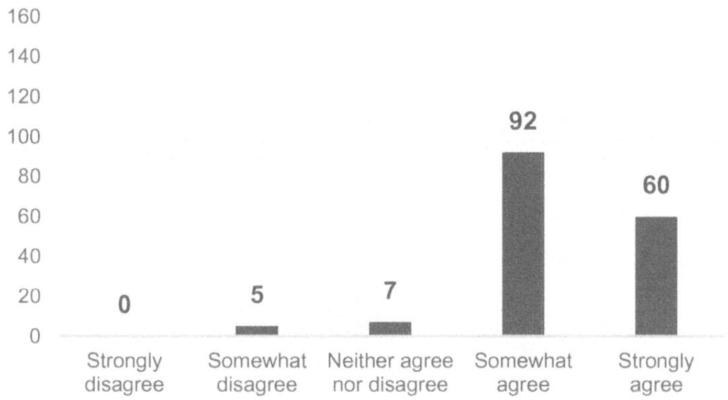

Fig. 4 Frequencies of responses to question 10 "**I practice eco-gestures in my daily life**" in the quantitative questionnaire on representations of EDM and smart meters

We note that 152 of the 164 respondents take daily energy-saving measures indicating an action-oriented attitude (Fig. 4).

These daily energy measures are certainly not impacting their comfort since they mostly do not find it normal to reduce comfort to reduce greenhouse gas emissions (Fig. 5). Shove (2012) explains the constant search for comfort is "*not only based on rational behaviors, but also depends on traditions, the internal management of the home, the relationship between sexes, and power relations within the family, which influence the environment differently*".

People are generally aware of the efforts needed to reduce the causes and effects of climate change. Unlike the results of the qualitative survey, the respondents trust technology to achieve energy targets (e.g. Fig. 6; 107 out of 164 respondents are in favor of technology as a solution to reduce energy consumption).

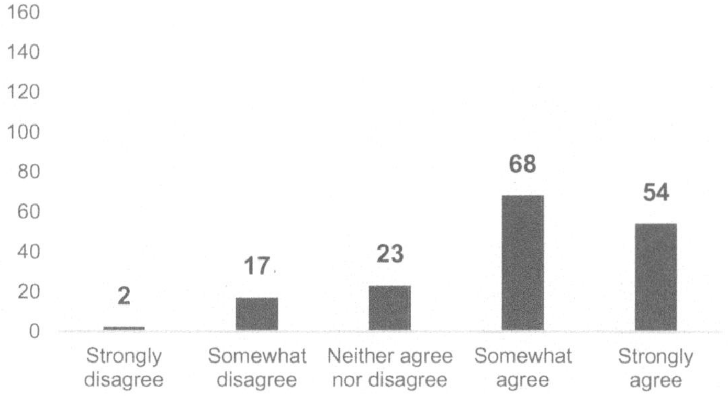

Fig. 5 Frequencies of responses to question 22 "**It is not normal to reduce comfort to reduce greenhouse gas emissions**" in the quantitative questionnaire on representations of EDM and smart meters

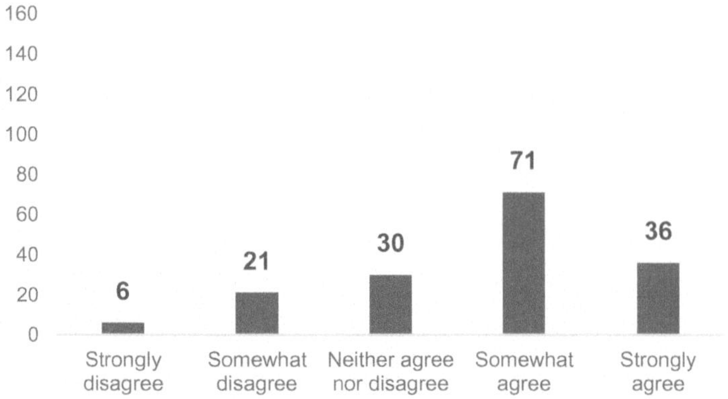

Fig. 6 Frequencies of answers to question 11 "**Technology will help us find solutions for the reduction of energy consumption**" in the quantitative questionnaire on representations of EDM and smart meters

Respondents expect efficient EDM tools, easy to use. A predictor of the acceptance of such a technical tool is the perceived usefulness. Data privacy is still an obstacle, and control over personal data remains a controversial topic (Fig. 7).

While 78 respondents felt that they do not have time to review their energy choices (Fig. 8), 118 respondents said that they would also like to have more information on the evolution of the energy market (Fig. 9).

We note that the economic factor would not be a determining factor in their energy choice: 96 respondents are ready to pay more to benefit from renewable alternative energy (Fig. 10).

However, fewer respondents (54, mostly managers) say that they do not pay attention to their energy consumption for financial reasons (Fig. 11). These quantitative results corroborate the results of the qualitative survey. The 50 respondents who say

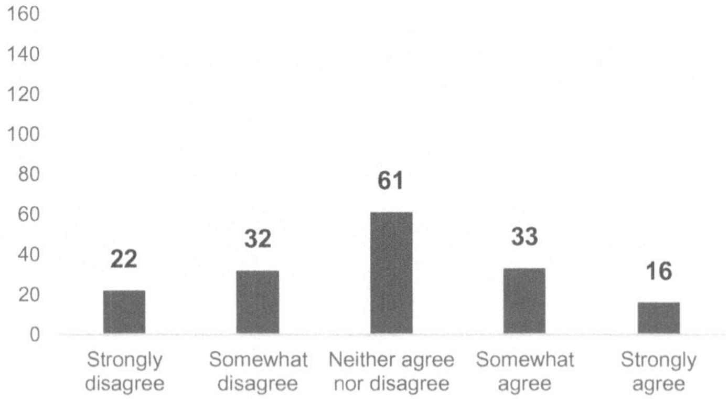

Fig. 7 Frequencies of responses to question 19 "**Smart meters infringe on privacy**" in the quantitative questionnaire on representations of the EDM and smart meters

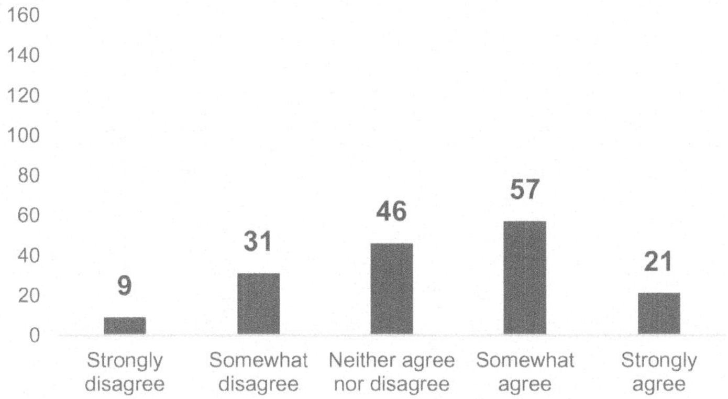

Fig. 8 Frequencies of responses to question 14 "**I don't have time to be an actor in my energy choices**" in the quantitative questionnaire on representations of DSM and smart meters

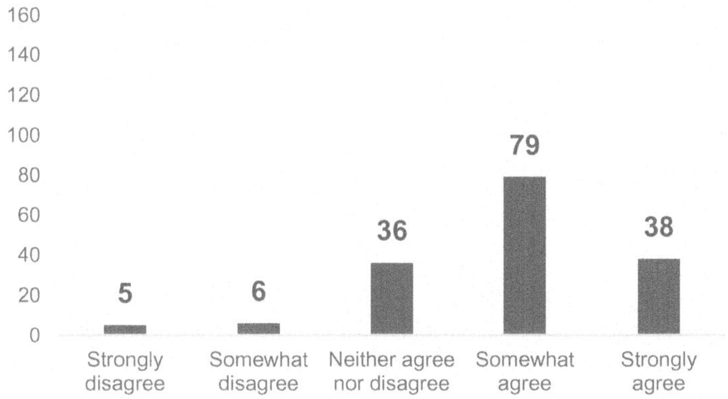

Fig. 9 Frequencies of responses to question 13 "**I would like to have more information on the evolution of the energy market**" in the quantitative questionnaire on representations of DSM and smart meters

that they pay attention to energy mainly for economic reasons have lower ecological reflexivity.

4 A New Tool for Ecological Reflexivity

The surveys carried out highlight a certain mismatch between the EDM tools offered and the consumers' expectations. The consumer is aware of the efforts to produce to reduce the causes and effects of climate change and is ready to act. They are quite confident in technologies, even if they have doubts about their efficiency. The doubts are reinforced by distrust in politicians and energy companies. On the other hand,

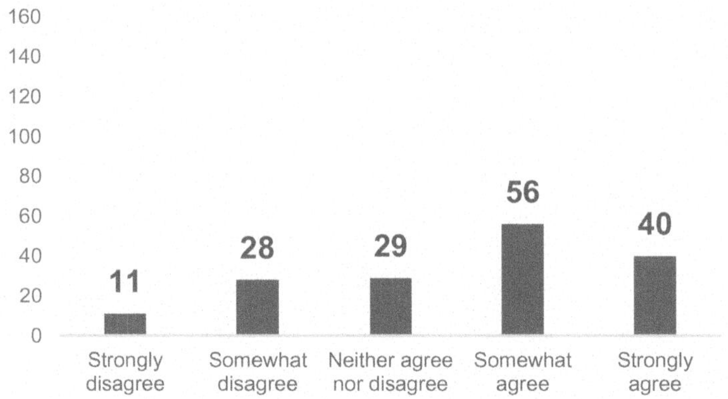

Fig. 10 Frequencies of responses to question 7 "**I am willing to pay more for alternative renewable energy**" in the quantitative questionnaire on representations of DSM and smart meters

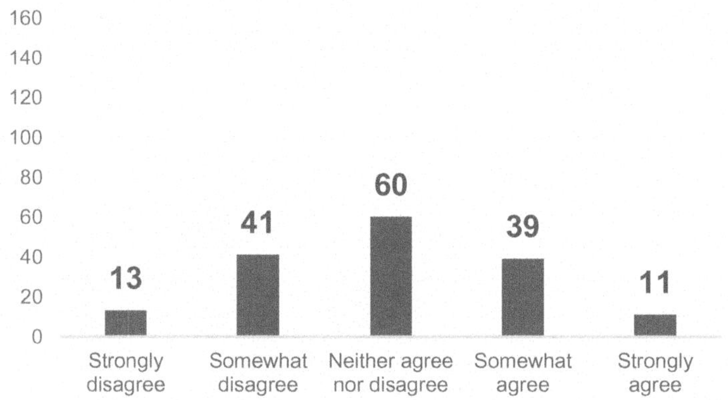

Fig. 11 Frequencies of answers to question 12 "**If I pay attention to electricity expenses at home, it is mainly for financial reasons**" in the quantitative questionnaire on representations of DSM and smart meters

actual EDM tools do not provide a comprehensive view of the objectives followed in the context of the energy and ecological transitions.

Providing consumers with better information on the overall societal issues and objectives, highlighting their lifestyle impacts on their energy consumption, and guiding them towards the most energy-efficient and ecological strategies would engage more and more people towards energy savings with increasing ecological reflexivity. Transparent information is needed to prevent consumers from elaborating false theories leading to a distorted representation of how they consume and how they could reduce their energy consumption (Kempton 1986; Kempton and Montgomery 1982; Krishnamurti et al. 2013). Lesic et al. (2018) analyzed 39 studies focusing mainly on individuals with higher ecological reflexivity to draw out insights into individuals' perceptions of energy consumption and savings. They show that individuals tend to overestimate the electricity consumption of low-energy devices, that they frequently use, and underestimate the high-energy devices that they use less frequently. The authors explain that these observations fluctuate with the unit used: overestimates are greater when using data in kWh rather than Wh. This unit effect has been observed in research on fuel by Allcott (2011), Larrick and Soll (2008); and water by Attari (2014). They also show that energy consumption reductions are generally preferred to energy efficiency strategies: consumers with the lowest ecological reflexivity found energy-saving measures, such as turning off lights and lowering the thermostat in winter, more effective than measures leading to improve the insulation of their home. Schwarz et al. (1991) explains that individuals judge the probability of an event according to the speed of emergence of an example in mind.

To facilitate individuals' access to information and improve their ecological reflexivity (a determining factor in decisions and action), a web interface is developed for the SMI tool. This development is being conducted in such a way as to overcome the obstacles that have been highlighted by our surveys. It must meet four main objectives:

- Promote awareness of energy and ecological issues (increase the level of ecological reflexivity);
- Engage users in monitoring energy consumption daily;
- Engage users in reducing their energy consumption with an ecological and not necessarily economic goal;
- Move towards sustainable behavior.

To encompass these objectives, the interface developed for the SMI tool is composed of three main modules (Fig. 12).

The first module (SMI) is dedicated to the analysis of data directly received from the SMI tool. This module gives the user the possibility to follow his electricity consumption (Which intensity? When do peaks in consumption occur? Which appliances consume the most? etc.) in several time slots. Several types of dynamic and easy-to-access displays are proposed. An "expert" display details the load curve, power peaks, daily consumption variations, average daily consumption curves by equipment, etc. Concerning the TTM model (Sect. 2.3; Fig. 2), this module provides the consumer with elements that must make him aware of the consequences of his energy consumption (upwards or downwards) of his or her practices. Consumption per equipment allows the user to identify equipment consuming more energy. The interface displays the data collected transparently, as trusting the tool and accepting to participate in reducing his/her consumption is an important outcome for the consumer.

The second "Do It Yourself" module is open to any user (whether they have the SMI tool or not). Users register their electricity consumptions (if they do not have an SMI) and other consumptions (gas, petrol/diesel, water, etc.) whenever they wish. About the TTM model, this second module allows users to act and collect their data. Making the consumer actively participate by mobilizing his knowledge and producing effort should favor the control of their consumption.

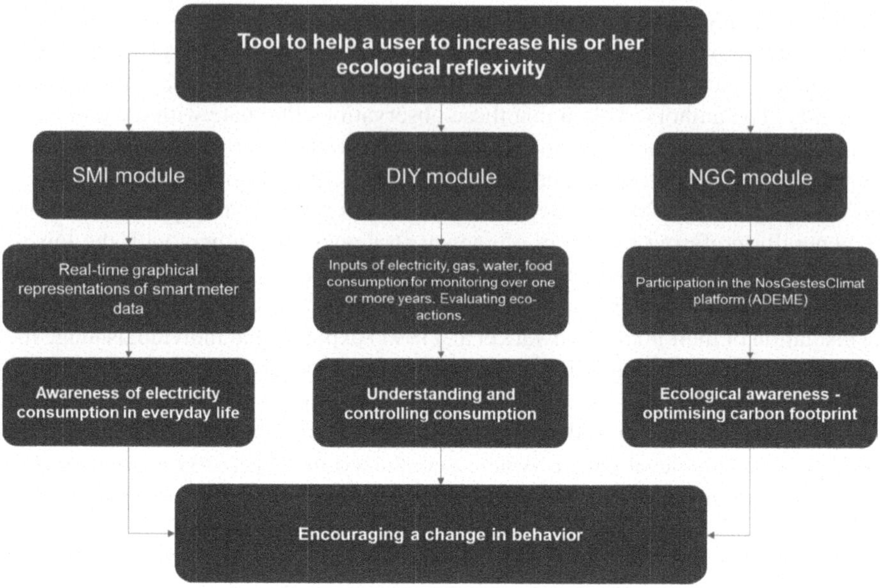

Fig. 12 Programmatic diagram of the web interface for the SMI project. *Source* Authors

The third and final module uses the free and open access module "Our Climate Gestures" (Nos Gestes Climat—NGC) available via the website of the same name. This NGC tool provides an example of an easy-to-use application, associated with transparent documentation of the underlying methodology. It also facilitates the transmission of data and the sharing of experiences between users. The module is based on an analysis of the user's annual carbon footprint, both overall and detailed by activity (food, transport, housing, digital, public services, others). It allows users to evaluate their carbon footprint and compare them to climate objectives. Ecological actions are proposed. Concerning the TTM model, this module mobilizes game elements motivating efforts towards new energy-saving and greenhouse gas emission reduction.

The three modules together allow users to analyze the data associated with their practices, using the simplest possible graphical representations, to test strategies for reducing their footprint, and share their data or experiences with other households. The display of economic gains associated with a reduction in energy demand and carbon emissions recontextualizes the participation of the users in the resolution of major global issues.

5 Conclusion

The combined analysis of the EDM tools and the relationship of users with these tools, and the EDM in general, allows us to identify a few obstacles.

According to our analysis, the actual EDM tools do not yet offer users to understand that their lifestyle influences a lot of their energy consumption and footprint. Respondents are mostly aware of the efforts to develop to succeed in the energy and ecological transitions and are ready to act. It should be noted that they have the most confidence in the technology, despite concerns about the security of the tools and the protection of privacy. Dobré (2003) and Subrémon (2013) mention a "*voluntary and thoughtful*" resistance, which refers to a desire for non-conformity on the part of users who do not want to be trapped in a harmful system that restricts part of their freedom. Taking back control of one's actions by putting up resistance against the automation of the home is also put forward by de Certeau, who evokes the act of 'doing oneself' (de Certeau 2010, reedition 1998). Subrémon (2009) explains that a set of practices implemented in the home would make it possible to promote 'energetic intelligence'.

EDM users monitor and save energy much more for economic reasons than for environmental protection. They weakly associate the EDM tools with the energy and ecological transition objectives. They are waiting for actions from the politicians and energy suppliers to progress in the energy and ecological transitions but do not trust their decision-making. They feel mostly disarmed face climate change (that is not here considered a controversial topic). Subrémon (2013) explains that the discourse on the ecological transition questions our lifestyles while a series of economic crises already generate fears and numerous other doubts among the households (work, family). Bovay et al. (1987) mentions that the social discomfort caused by all these uncertainties led households to favor family comfort. Our surveys highlighted that energy-saving measures are also not a priority in front of future uncertainty and may be neglected due to lack of time, disinterest, skepticism or even more general mistrust of policies.

The analysis of existing tools showed that energy and environmental issues are also poorly explained upstream of incentives to reduce energy consumption. This is perhaps a consequence of the ambiguous position of energy suppliers who are supposed to participate in the reduction of energy consumption while their profits depend on it.

Based on a literature review and surveys, we presented a web interface to the SMI tool. The aim is to propose an interface that can help users to understand the challenges of the energy and ecological transition and their consumption by associating it with the influence of their environment and lifestyle. As the analysis of surveys did not reveal specific consumer representation profiles, which could have been used to adapt the interface according to the user, the interface is one addressed to all users. The intention is to guide the evolution of users towards sustainable practices in different stages, and gradually increase their ecological reflexivity (i.e. to engage them more generally and in a sustainable manner in an awareness of the impact of their practices on the environment).

References

Allcott H (2011) Consumers' perceptions and misperceptions of energy costs. Am Econ Rev 101:98–104. https://doi.org/10.1257/aer.101.3.98
AlSkaif T, Lampropoulos I, van den Broek M, van Sark W (2018) Gamification-based framework for the engagement of residential customers in energy applications. Energy Res Soc Sci 44:187–195. https://doi.org/10.1016/j.erss.2018.04.043
Attari SZ (2014) Perceptions of water use. Proc Natl Acad Sci USA 111:5129–5134. https://doi.org/10.1073/pnas.1316402111
Beck AL et al (2019) Not so gameful: a critical review of gamification in mobile energy applications. Energy Res Soc Sci 51:32–39
Boone HN, Boone DA (2012) Analyzing Likert data. J Extens 50:1–5
Bovay C, Campiche R, Hainard F, Kaiser H, Pedrazzini J, Ruh H, Spescha P (1987) Energie au quotidien. Labor et Fidès
Brisepierre G (2011) Les conditions sociales et organisationnelles du changement des pratiques de consommation d'énergie dans l'habitat collectif
Brounen D, Kok N, Quigley JM (2012) Residential energy use and conservation: economics and demographics. Eur Econ Rev 56:931–945
Cayla J-M, Allibe B, Laurent M-H (2010) From practices to behaviors: estimating the impact of household behavior on space heating energy consumption. In: ACEEE summer study on energy efficiency in buildings
Comby J-B, Grossetête M (2012) «Se montrer prévoyant»: une norme sociale diversement appropriée. Sociologie 3(3):251. https://doi.org/10.3917/socio.033.0251
de Certeau M, Giard L, de Certeau M (2010, re-edition 1998) Arts de faire. In: de Certeau M (ed) L' invention du quotidien, Nouvelle éd. Gallimard, Paris
Dobré M (2003) L'écologie au quotidien: éléments pour une théorie sociologique de la résistance ordinaire, Sociologies et environnement. l'Harmattan, Paris, Budapest, Torino
Druckman A, Jackson T (2008) Household energy consumption in the UK: a highly geographically and socio-economically disaggregated model. Energy Policy 36:3177–3192
Ehrhardt-Martinez K, Donnelly KA, Laitner S (2010) Advanced metering initiatives and residential feedback programs: a meta-review for household electricity-saving opportunities. American Council for an Energy-Efficient Economy, Washington, DC
Gram-Hanssen K, Bech-Danielsen C (2004) House, home and identity from a consumption perspective. Hous Theory Soc 21:17–26
Herpin N, Verger D (2008) Consommation et modes de vie en France: une approche économique et sociologique sur un demi-siècle. Découverte
Kempton W (1986) Two theories of home heat control. Cogn Sci 10:75–90. https://doi.org/10.1207/s15516709cog1001_3
Kempton W, Montgomery L (1982) Folk quantification of energy. Energy 7:817–827. https://doi.org/10.1016/0360-5442(82)90030-5
Kendel A (2015) Smart Meter, Feedback et maîtrise de la consommation électrique: Le cas du secteur résidentiel dans la commune de Biot-Alpes Maritimes. Sciences Economiques, Université de Nice-Sophia Antipolis, Nice, 217 p
Krishnamurti T, Davis AL, Wong-Parodi G, Wang J, Canfield C (2013) Creating an in-home display: experimental evidence and guidelines for design. Appl Energy 108:448–458. https://doi.org/10.1016/j.apenergy.2013.03.048

Larrick RP, Soll JB (2008) The MPG illusion. Science 320:1593–1594. https://doi.org/10.1126/science.1154983
Lesic V, de Bruin WB, Davis MC, Krishnamurti T, Azevedo IML (2018) Consumers' perceptions of energy use and energy savings: a literature review. Environ Res Lett 13:033004. https://doi.org/10.1088/1748-9326/aaab92
Lutzenhiser L (1993) Social and behavioral aspects of energy use. Annu Rev Energy Environ 18:247–289
Lutzenhiser L, Gossard MH (2000) Lifestyle, status and energy consumption. In: Proceedings of the 2000 ACEEE summer study of energy efficiency in buildings, pp 8–207
Maresca B, Dujin A, Picard R (2009) La consommation d'énergie dans l'habitat entre recherche de confort et impératif écologique. Centre de recherche pour l'étude et l'observation des conditions de vie
Prochaska JO, Velicer WF (1997) The transtheoretical model of health behavior change. Am J Health Promot 12:38–48. https://doi.org/10.4278/0890-1171-12.1.38
Roy A (2007) Les pratiques environnementales des Français. In: Problemes politiques et sociaux, p 97
Sahakian M (2011) Understanding household energy consumption patterns: when "West Is Best" in Metro Manila. Energy Policy 39:596–602
Sanquist TF, Orr H, Shui B, Bittner AC (2012) Lifestyle factors in U.S. residential electricity consumption. Energy Policy 42:354–364. https://doi.org/10.1016/j.enpol.2011.11.092
Schwarz N, Bless H, Strack F, Klumpp G et al (1991) Ease of retrieval as information: another look at the availability heuristic. J Pers Soc Psychol 61:195–202. https://doi.org/10.1037/0022-3514.61.2.195
Schweitzer V, Simon F (2021) Représentations sociales paradoxales des compteurs Linky et paradoxe de la vie privée: une lecture à partir du cadre théorique de l'empowerment psychologique 13
Shove E, Pantzar M, Watson M (2012) The dynamics of social practice: everyday life and how it changes. SAGE Publications Ltd. https://doi.org/10.4135/9781446250655
Subrémon H (2009) Habiter avec l'énergie: Pour une anthropologie sensible de la consommation d'énergie. Université Paris X, Nanterre
Subrémon H (2013) Habitudes de consommation d'énergie des ménages: état des lieux 10
Tabbone L (2017) Consommations énergétiques et cadres de vie: analyses en termes de modes de vie. https://doi.org/10.5075/EPFL-THESIS-8045
Zélem M-C (2010) Politiques de maîtrise de la demande d'énergie et résistances au changement: une approche socio-anthropologique, Logiques sociales. l'Harmattan, Paris
Zélem M (2018) Économies d'énergie: le bâtiment confronté à ses occupants. Annales des Mines - Responsabilité et environnement 90:26–34. https://doi.org/10.3917/re1.090.0026

Security Aspects of Smart Meter Infrastructures

Ivan Rigoev and Axel Sikora

1 Introduction

One of the most important questions about smart metering systems for the end users is their data privacy and security. Indeed, smart metering systems provide a lot of advantages for distribution system operators (DSO), but functionalities offered to users of existing smart meters are still limited and society is becoming increasingly critical. Smart metering systems are accused of interfering with personal rights and privacy, providing unclear tariff regulations which not sufficiently encourage households to manage their electricity consumption in advance. In the specific field of smart grids, data security appears to be a necessary condition for consumer confidence without which they will not be able to give their consent to the collection and use of personal data concerning them.

From the personal data privacy and security side several articles are showing that with advanced power signature analysis tools such as Nonintrusive Appliance Load Monitoring (NIALM), attackers can determine the types and times of electrical appliances in a home, as well as learn detailed information about a resident's daily activities. Batra et al. (2014) have developed methods for determining the use of electrical appliances using "consumer profiling". Murrill et al. (2012) showed that analyzing energy consumption data over 15 min can determine the types and quantities of appliances used in a home. Molina-Markham et al. (2010) describe how a consumer can be "profiled" using only generic statistics, without detailed network signatures for electrical appliances, and without prior training. Greveler et al. (2012) showed that it is possible to identify which channel is watched on TV only by analyzing consumption data from smart meters. Also, some smart metering devices contain vulnerabilities that could be used by hackers. For example, security

I. Rigoev (✉) · A. Sikora
Institut für Verlässliche Embedded Systems und Kommunikationselektronik (ivESK), Hochschule Offenburg, Badstraße 24, 77652 Offenburg, Germany
e-mail: ivan.rigoev@hs-offenburg.de

D. Ould Abdeslam (ed.), *Smart Meters*, Lecture Notes in Energy 97,
https://doi.org/10.1007/978-3-031-27556-2_5

engineers in Spain discovered that local electricity company smart meters used the same key for AES128 encryption on all devices (Higgins 2014). Or discovering that no security measures are being used because of wrong settings (Carracedo 2019). Due to the significant development of intelligent networks and the data exchanged, hacking into networks or taking control of electrical infrastructures remotely could have dreadful consequences such as the paralysis of the electrical network.

To better understand the current state of French, German, and Swiss smart metering systems were made a comparative security analysis of these systems and it is presented in this chapter.

In Sect. 2 we will consider smart metering supervisory authorities, their requirements, and common smart metering solutions and compare them with reference smart metering infrastructure architecture model.

In Sect. 4, we will describe our methodology and provide more a detailed description of DLMS/COSEM protocol, its security, protocol and common implementations vulnerabilities, a comparison of DLMS/COSEM with TLS with BSI restrictions, and make a theoretical comparison of considered SMI systems.

In Sect. 6 we will present some parts of the penetration testing analysis of two BSI-certificated SMGW devices.

The last section contains some recommendations for future developments and already used smart metering systems.

2 Security Architectures—A Comparative Overview

In this section, we will summarize French, German, and Swiss legal acts and other available documentation related to smart metering. The objective is to describe their architectures and compare them with Smart Metering Coordination Group reference architecture for smart metering communications, which is described in the framework of the smart grids M/490 mandate (Sánchez Jiménez 2011) and smart meters M/441 mandate (CEN/CLC/ETSI/TR 50572, 2011).

2.1 French Case

In France, the protection of personal data confidentiality is guaranteed by the "Data Protection Act" 78-17 of 6 January 1978 "On Data Processing, Data Files, and Individual Liberties" (French National Assembly and the Senate 1978). This act has a wide scope of application, since all information relating to an identified or identifiable natural person, directly or indirectly, is considered as personal data. This act covers all the processing of this data and also gives a broad definition of the processing. Thus, the mere collection of personal information constitutes the processing of these data within the meaning of the law, as well as data conservation. Terms of the areas covered by this act—are very similar to the GDPR. Compliance with this act must be carried out under the control of the "National Commission of data processing and freedoms" (Commission Nationale de l'informatique et des libertés, CNIL).

In addition to a guarantee of data confidentiality, ensuring the absence of communication without consent by users, the securing of this data against malicious actions and piracy is provided by the regulations. The Data Protection Act required processing of personal data to "take all useful precautions, with regard to the nature of the data and the risks presented by the processing, to preserve the security of the data and, in particular, prevent them from being distorted, damaged, or from unauthorized third parties having access to them". This general obligation of security applies to all categories of personal data subject to processing.

Specifically, regarding electricity metering systems, Decree 2001-630 of 16 July 2001 (Légifrance 2001) [Decree 2004-183 of 18th February 2004 for gas (Légifrance 2004)] requires system operators to keep confidential commercially sensitive data (information whose disclosure could undermine the rules of free and fair competition and non-discrimination). Metering data are commercially sensitive. Since 2010, work has been carried out in France between the CNIL and the Energy Regulatory Commission (Commission de régulation de l'énergie, CRE) on the implementation of smart meters, focusing in particular on the question of data collected by this equipment processing.

In the European Commission recommendation from 9 March 2012 (EUR-Lex 2012) "on preparations for the roll-out of smart metering systems", European Commission has recommended that data protection should be integrated into the functionalities of equipment from their design and that their default settings should be as protective as possible of the security and confidentiality of personal information. For its part, intending to protect privacy, the CNIL has prohibited the collection of data that relates to consumption with a time step of fewer than 10 min for communicating electricity meters, which required avoiding too precise electricity consumption knowledge. General functionalities of French smart meters have been defined in the application of these CNIL recommendations.

Thus, the order of 4 January 2012 (Légifrance 2012a) defining the functionalities of smart electricity meters provides that consumption readings are taken at a time step that cannot be less than 10 min. However, this limitation only concerns installations connected at low voltage for power less than or equal to 36 kVA, which mainly corresponds to private installations, while installations connected at higher power may give rise to more regular readings. In response to the need for security, this order requires that the metering devices comply with safety standards approved by the Minister for Energy, this compliance being subject to verification and certification by the ANSSI. Concerning electricity metering, the order of 4th January 2012 (Légifrance 2012a) requires system operators to have their metering system certified under Decree 2002-535 of 18 April 2002 (Légifrance 2002).

The National Cybersecurity Agency of France (The Agence Nationale de la sécurité des systèmes d'information, ANSSI) is a French service created on 7 July 2009 with responsibility for cybersecurity (https://www.ssi.gouv.fr/en/mission/word-from-director-general/). ANSSI replaced the Central Directorate of Computer Security, which on July 31, 2001, replaced the Service central de la sécurité des systèmes d'informations (SCSSI Eng. Central Service for Information System Security). By Decree No. 2009-834 of 7 July 2009 (Légifrance 2009) as amended by Decree No. 2011-170 of 11 February 2011 (Légifrance 2011), the agency has responsibility at the national level concerning the defense and security of information systems. It is attached to the Secretariat-General for National Defence and Security (Secrétaire général de la défense et de la sécurité nationale) under the authority of the Prime Minister. ANSSI is responsible for proposing rules for the protection of state information systems and verifying the implementation of measures adopted. In the field of cyber defense, it provides a monitor, detects, alerts, and reactions to computer attacks, especially on state networks.

ANSII considers a certification scheme called CSPN (Certification de Sécurité de Premier Niveau; https://www.ssi.gouv.fr/administration/produits-certifies/cspn/), aims to provide a first-level security certification for IT security products. CSPN set up by ANSSI in 2008 consists of "black box" tests carried out under tight deadlines. Its scope is similar to the vulnerability analysis performed within Common Criteria, with the following specificities:

- The assurance process is simplified
- The evaluation is focused on vulnerability analysis
- The actors are committed to a given evaluation duration and cost.

IT products can currently apply to CSPN if they belong to a specific list of domains (e.g. data deletion, firewalls, secure communication, etc.). This list is regularly updated to address new needs. It should be noted that standard CSPN excludes products too complex to be evaluated in an expected duration and cost and products including non-standard cryptography. The organization and evaluation methodology process is similar to the Common Criteria process. Instead of applying CC security and assurance requirements, the developer uses guidelines described in CSPN. CSPN also has common features with CPA, especially the domain-specific approach. CSPN is used for meters and data concentrator security certification.

On 15 November 2012 (Légifrance 2012b), the CNIL adopted a recommendation that sets the framework and conditions under which consumption data from smart electricity meters may be collected and processed. CNIL recalled that the implementation of smart meters presents risks concerning privacy and recommended the introduction of strong technical measures guaranteeing data security. Beyond IT security, the CNIL has recommended that all players processing data collected by the network operator implement additional security measures because the increase in the level of detail of the data leads to an increase in the risk of invasion of privacy. In this recommendation, the CNIL recalls that the future deployment of smart meters will lead to a very large amount of data collected. In particular data relating to the quality of the electricity supplied to the subscriber or consumption indices, which are already used and processed data. Above all, smart meters offer a new functionality by making it possible to collect information relating to the load curve, consisting of reading at regular intervals of the subscriber's electricity consumption which, according to the CNIL, would make it possible to have information about people lifestyle details (identification of wake up and bedtime, periods of absence, even the volume of hot water consumed per day or the number of people present in the accommodation). The CNIL, therefore, wanted to regulate the conditions for collecting and use of this load curve by smart meters, recalling that these operations are subject to the Data Protection Act. On the other hand, the storage of the load curve by the meters, without the feedback of the information to the network manager, complied with the requirements of the Data Protection Act.

In general, the Data Protection Act requires the prior consent of the person concerned to be obtained before any processing of personal data, except under certain specific conditions (when the processing of this data is subject to a legal obligation, execution of a public service mission or a contract, for example). Data subjects have the right to access and rectify data concerning them and may object, for legitimate reasons, to the processing of their data. The Data Protection Act also requires that the automated processing of personal data being the subject of a declaration to the CNIL, or even of authorization from this commission for certain categories of data (biometric data, relating to offenses, etc.). The Data Protection Act subjects those responsible for processing personal data to several obligations in terms of informing the persons whose data is processed and regulating the duration of data retention which must be proportionate to the purpose for which the data is collected.

The Data Protection Act fully applies to data from smart grids and no additional legal instrument had to be adopted to guarantee the confidentiality of this data. However, since actors in the energy sector are less familiar with this regulation than in other sectors where data management is already well known problem, the CNIL specified in Deliberation 2012-404 of 15 November 2012 (Légifrance 2012b) how this is applied to the specific area of smart grids, primarily on data collected (consent and limiting load curve sampling period), the duration of data retention (no conservation beyond the time required), the recipients of the data (habilitation) and security measures (assessment and regular updating).

The CNIL has indicated that the processing of the load curve can only be implemented for three purposes:

- maintenance and development of the network (by the DSOs)
- establishment of tariffs adapted to household consumption (by energy suppliers)
- provision of complementary services (by third-party companies).

Taking up the general obligation provided in Article 6 of the Data Protection Act, the CNIL recommended limiting the data retention period to the time necessary to achieve the purpose for which these data are collected. Compliance with this obligation is monitored by the CNIL, which can impose pecuniary sanctions after formal notice to the persons concerned. Penal sanctions are also incurred, ranging up to five years of imprisonment and a fine of 300,000 euros.

In addition, it specifies that the framework for the collection and use of the load curve must be defined according to the recipient of this collection:

- when the recipient is the network manager, the collection of the load curve must be carried out only when power supply problems are detected and not systematically;

- when the recipients are suppliers and service providers, and the load curve is used to set tariffs adapted to household consumption and the supply of additional services, it is necessary to collect the load curve beforehand. Express consent of the persons concerned, which must be free, informed, and specific.

CNIL has developed, within the framework of a partnership with the Fédération des Industries Electriques, Electroniques et de Communication (FIEEC), a "Compliance pack for communicating meters" (Pack de conformité compteurs communicants) (CNIL 2014) which is presented as a guide to good practice intended for manufacturers giving concrete indications on how to comply with the texts relating to the protection of personal data, and specific operating methods.

This compliance pack provides recommendations on how to handle data collected by smart devices based on three scenarios:

- the "IN-IN" scenario, which concerns the management of data collected in the home without communication with the outside (for example, communication mechanisms between the thermostat and the heating or smartphone applications that only send information to the user). This scenario does not impose any formalities before data collection but requires security measures to ensure that no unauthorized person has access to the data;
- the "IN-OUT" scenario, which applies to the management of data collected in the dwelling and transmitted outside (for example, electricity consumption data transmitted to a third party for a thermal renovation service). The service provider must make a declaration to the CNIL and must obtain the prior consent of the person concerned;
- the "IN-OUT-IN" scenario, which concerns the management of data collected in the dwelling and transmitted outside to allow remote control of certain equipment in the dwelling (for example, remote production control system domestic hot water). As for the previous scenario, the service provider must make a declaration to the CNIL and must obtain the prior consent of the person concerned.

Regarding data security technical aspects the Compliance pack has only very abstract requirements which are not so much different for different scenarios. For example, the security part from CNIL Scenario 2 "IN–OUT" contains such requirements as:

- The service provider shall put in place measures to ensure the security and confidentiality of the data processed by the devices provided to the person and shall take all appropriate precautions to prevent being taken over by an unauthorized person, in particular by:
- Providing encrypted data exchanges with state-of-the-art algorithms, protecting the encryption keys from any accidental disclosure, authenticating the devices receiving the data, and making access to the control functions of the installation conditional on authentication of a reliable user (password, electronic certificate, etc.).
- Thus, the measures put in place must be adapted to the level of sensitivity of the data. Regarding the measures to be put in place at the level of infrastructures

external to the housing, the service provider must carry out a study of the risks generated by the treatment to determine and implement the measures necessary to protect the privacy of individuals.

Thus, the application to the energy sector of the regulations relating to data protection should make it possible to prevent the breaches of privacy that the development of the digitalization of networks could have risked causing. Although the documentation produced by the CNIL (recommendations and compliance pack) does not in itself have a binding legal force, it helps guiding professionals by illustrating the concrete application to smart networks of the Data Protection Act.

Because in France smart grids are considered as specific Industrial Control Systems—the most appropriate document with requirements from ANSII is "Cybersecurity for Industrial Control Systems—Detailed Measures" from January 2014 (ANSSI 2014). Requirements from this document theoretically should be applicable for smart metering systems because the working group was not interested in a particular industry and the elements contained in this document are therefore intended to apply to all sectors. Security requirements in this document are more concrete, but it is not fully clear which of these requirements should be applied for smart metering systems.

The importance and usefulness of carrying out impact studies before the deployment of smart grid equipment, following the European recommendations which are described above, was affirmed by CRE in the deliberation of 12 June 2014 (CRE 2014) providing recommendations on the development of low voltage electrical networks, although this is not a legal obligation.

One more important document is the French "Energy Code" (Code de l'énergie). The Energy Code is an official French legal code bringing together various provisions relating to energy law. Energy Code is a very large document, but the most important parts of the document are:

- DSOs are responsible for Energy Individual Data Protection.
- DSOs must ensure access to individual customers to their own energy data via a secured web portal.
- Individual customers must also be able to share—or authorize DSOs to share their data with any authorized third party (through express customer consent).

However, the consent process for individual customers is not yet fully defined (in particular roles and responsibilities of DSOs and Providers/Third Parties). Only in some specific cases, such as for "Flexibility aggregators" and "Energy Providers", there are clear indications of the need of collecting customer consent before getting any data from DSOs. There is no clear indication concerning the process of controlling consents collected on the provider side by the DSO and this is perceived as a weakness because there is no clear guarantee that customer data is properly used.

Enedis and GRDF

All documentation described above does not contain architecture or communication protocol requirements for smart metering systems. There are about 144 distribution system operators in France (Rullaud and Gruber 2020), but Enedis and GRDF dominate the electricity and gas DSO market. The problem with the lack of information in French documentation for comparison is partly solved because both Enedis and GRDF have only one possible solution for smart metering—Linky smart meter for electricity and Gazpar for gas. The following materials about French smart metering solutions in this document will refer to these systems.

Gaz Réseau Distribution France (GRDF) is a French gas distribution company founded on January 1, 2008. It is the main distributor of natural gas in France and Europe. It is a 100% subsidiary of Engie. GRDF took over the activities previously carried out by EDF Gaz de France Distribution, which operated as a directorate of Gaz de France with independent management (https://www.grdf.fr/english/leading-natural-gas-distribution-operator). Widespread deployment of the advanced Gazpar meter was approved by a decision of the Minister of Ecology, Sustainable Development and Energy and the Minister of Economy, Industry, and Digital on 23 September 2014 (Ministère de l'Écologie DDDEDL 2014). Gazpar should be deployed in 11 million homes by 2022 (https://www.grdf.fr/institutionnel/actualite/dossiers/compteur-communicant-gazpar). As a Linky meter, Gazpar allows remote reading and transmission of actual consumption indices.

Enedis, formerly ERDF (for Électricité Réseau Distribution France), is a public limited company with a supervisory board and management board, a wholly-owned subsidiary of EDF responsible for managing and developing 95% of the electricity distribution network in mainland France (https://www.enedis.fr/qui-sommes-nous). Enedis was created on 1 January 2008, under the name ERDF, by splitting EDF's electricity distribution activities into electricity generation, transmission, and marketing activities.

The first brick in the rollout of smart grids in the French electricity sector was the deployment of the Linky smart electricity meter which was decided by Law 2005-781 of 13 July 2005 (Légifrance 2005) on the orientations of the energy policy. This communication meter can collect consumption data (quantity consumed per hourly interval or power requested by the consumer) and can provide information to the customer himself, in particular to equipment inside the household, or third parties such as energy service providers (via Solenn). The download of daily data from the Enedis portal is possible at any time for customers in a simple standard file format. However, Enedis is working on an alternative file format, based on the IEC international data model CIM, which could be used as a starting base for the next steps in this ad hoc group to build a common interoperable standard. Collecting half-hourly data is possible only for customers that have provided their consent. Enedis may have access to half-hourly data without collecting consent only for operating needs, in a limited timeframe, and in a specific area. Besides, consumers can also directly access their consumption data in near real-time via special adapter devices connected to the Linky meter.

Sharing data through the Enedis data exchange platform with a third party is currently possible only with energy providers, and it is not generalized yet. Separate consent is requested for data transmission and use but is not necessarily collected by the Enedis. The consent mechanism works as follows: customers equipped with a Linky smart meter are invited to connect to the Enedis web portal where they can consent to the collection of their half-hourly data, storing in the information system, and downloading. Customers can easily opt-out of this process and so far, the web portal does not include the possibility to provide consent to share data with third parties. Providers can send requests to Enedis for starting load curve data collection (monthly to daily data and if available half-hourly data) and accessing customer data. There is no requirement to show the customer consent upfront, but the provider must be able to provide it if asked, to get access to daily and infra-daily data.

To limit the risk of intrusion into privacy, CNIL has strictly supervised data collection. Linky smart meters only report information concerning the consumer's total consumption and do not allow specific uses to be distinguished. The Linky meter can measure three main types of data (UFC-Que Choisir 2017):

- Consumption indexes. Before, they were estimated or transmitted either by the distributor to the supplier, or by the consumer, to establish the invoicing. Now they can be reassembled automatically. No more than before, this information will not allow the distributor and suppliers to know consumption habits.
- The load curve, i.e. the graphic representation of the evolution of energy consumption over a given period. It consists of reading the subscriber's electricity consumption at regular intervals (the time step). This data could pose a problem because it would then be possible to determine at what time of day consumption is more or less important. To stem this risk, the CNIL has imposed that the transmission of the load curve is explicitly consented to by the consumer. In addition, if agreed, the interval at which data is uploaded to Enedis cannot be less than 10 min. Below this period, it is indeed possible to identify the uses that the consumer makes of his devices.
- Data relating to meter quality and security. These data are not personal. They allow Enedis to check the quality of the power supply, power cuts, or even check the openings of the meter cover to prevent fraudulent acts. The collection of this data does not make it possible to know the consumption habits of the consumer.

Refusing to allow Enedis access to its meter to replace it with Linky is illegal (https://www.fournisseurs-electricite.com/guides/compteur/linky/refuser) insofar as electricity metering devices are the property of the local authorities, which grant Enedis management of them, as provided for in Article L322-8 of the Energy Code. The same article also specifies that the DSO is responsible for intervening on the meters for operations such as connection, commissioning, and shutdown of meters or power changes.

Electricity distribution in France is a public service. In accordance with the provisions of concession contracts concluded between local authorities and the network manager, the latter is responsible for the execution of this public service, which he must ensure in compliance with the law and the regulation. However, the law requires the implementation of counting devices. By opposing the installation of Linky meters, clients take the risk of opposing the execution of a public service mission.

In addition, when clients conclude an electricity supply contract, they join the general provisions relating to the access and use of the public distribution network

(GRD contract). In its 2016 version, this contract indicates that (UFC-Que Choisir 2017):

- The customer must commit to "taking any provision to allow DSO to carry out the installation, modification, maintenance, and verification of counting equipment" (art. 2.3).
- The Customer is responsible for "direct damage and some caused in DSO in the event of non-compliance with one or more of the obligations charged to him for access and use of the RPD [Public Network for the Distribution of Distribution electricity, editor's note]" (art. 6.2).
- DSO may make the suspension or refuse access to the RPD, in particular in the event of "non-justification of the compliance of the installations to the regulations and the standards in force" (art. 5-5, point 5).

However, the law requires the implementation of smart meters. Concretely, this means that:

- If a consumer will not allow DSO to make the installation or modification of the counting equipment, DSO will be deprived of the possibility of carrying out a remote meter statement and will therefore be founded to invoice the consumer a special statement.
- By refusing DSO the installation of the meter, the consumer would refuse to bring it up to standards and therefore be exposed to the suspension of access and use of the RPD.

If the installation of the meter has been refused by the consumer, CRE admits that the succession of the meters is billed by the network manager thus causing additional costs for the user (CRE 2016).

Enedis Smart Metering Architecture

Enedis smart metering architecture and used protocols could be founded in the document Linky PLC profile specifications (ERDF-CPT-Linky-SPEC-FONC-CPL) which for some reason currently unavailable on the Enedis website, and only version 1.0 from the year 2009 possible to find on the internet (EDRF 2009). This document describes a stack of protocols that should be used for the Linky system including the application layer protocols. On top of PLC should be used Logical Link Control protocol (LLC) IEC 61334-4-32, IEC 61334-4-41:1996 Distribution automation using distribution line carrier systems—Part 4: Data communication protocols—Section 41: Application protocol—Device Language Message Specification (DLMS), and COSEM from Blue and green "Coloured Books". The objects associated with PLC network management are defined by the COSEM class instances ID 50, 51, 52, 53, and 56 described in the Blue Book. The encoding rules are described in the A-XDR standardization document—Distribution automation using distribution line carrier systems—Part 6: A-XDR encoding rules. The objects accessed via the COSEM application layer are identified according to the rules specified in IEC62056-61 Electricity metering—Data exchange for meter reading, tariff, and load control—Part 61: OBIS Object identification.

Document IDIS (Interoperable Device Interface Specification) White Paper points out that Linky smart meter meets IDIS specifications, and ERDF has issued the LINKY companion specifications describing how these standards are used and which options are chosen (but we were not able to find this document). IDIS specifications will be considered more precise in Sect. 4.3.

Enedis smart metering architecture includes a Linky meter that is commonly installed inside a private territory, a data concentrator installed at non-private territory, and supervisory control servers (Fig. 1). Data concentrator units are used to interrogate meters, to process and store the information it receives, and to send this data to a centralized Information System. Enedis concentrators do not decrypt data (only resend it to a backend system). Every Linky meter connected to a Data concentrator device via PLC (power line carrier). By default, data concentrators are connected to Enedis servers via a cellular network. Common Enedis protocols stack has no other options for these connections.

PLC—is a protocol that uses the possibility to transfer data using common power grid communications by adding high-frequency electrical signals over main 50 Hz (Chauvenet 2016). The main advantage of this technology is that additional communication wires of telecom infrastructure do not require, and that allows to be independent of telecom operators. The most significant shortcoming of PLC technology is that PLC can produce a good signal only at approximately 100 m. To solve this problem—every Linky meter works like a mesh node (after receiving a signal Linky repeats it). On average Enedis use one concentrator per 60 Linky meters, but 1 concentrator can cover up to 1000 smart meters with PLC-G3.

Additionally, the Linky system provides a list of useful features, such as:

- Recording and remote transmission of supply-quality data
- Remote home appliance control
- Energy box management

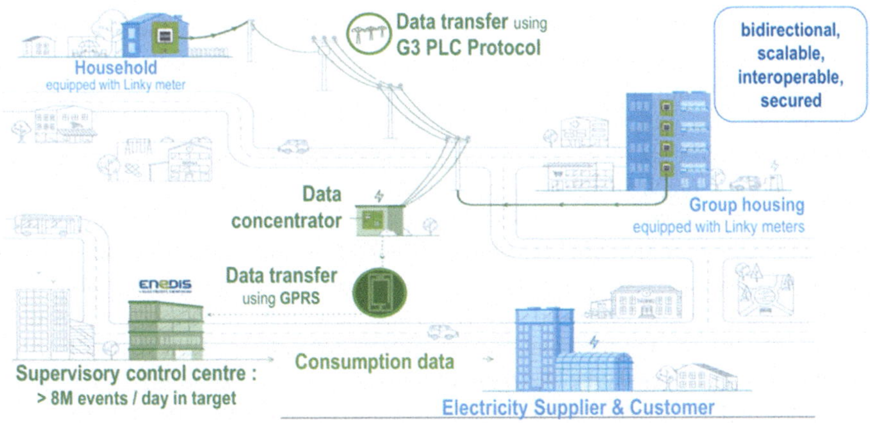

Fig. 1 Enedis DSO smart metering architecture (SMARTER TOGETHER 2019)

- Remote control of the data concentrator and diagnostic
- Remote interrogation of a meter group (Grouped Meter Ping)
- Power outage detection
- Analysis of the loss of power supply on the data concentrator
- Loss of phase alarms on triphased meters
- Reverse voltage detection alarm on the data concentrator
- Remote transmissions of fault detectors connected to the data concentrator
- Analysis of PLC connection between meters and their data concentrator
- "Predictive Maintenance" tools such as CartoLine.

One of the Linky system features is the ability to manage the power consumption of certain appliances in the home. Enedis does not have direct access to the power of home appliances and can only give a binary signal to the "advice" to turn off some of the 7 virtual ports. Physically Linky has only one port—dry contact (French name "Contact sec") (enedis.fr Notice d'utilisation du compteur communicant Linky).

Energy box is a battery device that allows storing energy that could be used in a time when energy cost is high. Linky can count both consumed energy and generated energy (e.g., by private solar panels). When solar panels generate more energy than consumed energy, it can be returned to the energy system to make a profit.

CartoLine is a "Predictive Maintenance" tool for Enedis LV networks. The question answered by CartoLine can simply be summed up as: does the observed voltage data require the intervention of a field technician to avoid a breakdown? The CartoLine tool, developed by Enedis R&D, uses the mass of data relating to the voltages observed by Linky meters to significantly improve the management of predictive maintenance on the low-voltage network. For experts, it offers a Dataviz interface facilitating data analysis. Above all, artificial intelligence runs these analyses in "supervised learning" first and then can conduct them autonomously. These analyses aim to identify situations reflecting a voltage anomaly or a future incident that could lead to a power outage on the LV network.

About security—Enedis mentioned that the security of the Linky system builds on a chain that goes from the meter itself to the Enedis data centers, via the thousands of concentrators deployed throughout France. At each stage, different points of attention were addressed (Marcellin 2018).

About the Linky system field part—each meter is CSPN certified. It has a key to encrypt (in AES standard, symmetric) the retrieved data locally and send it to the concentrator (each of these groups together 10–10,000 m, depending on the concerned area). A concentrator was designed as a digital safe device and it also CSPN and "Common Criteria" certified on behalf of the hardware. The data concentrator deletes the received data in the event of an intrusion detection or even abnormal operation. Another encryption (asymmetric based on elliptic curves) is used between the concentrator and the information system (IS) dedicated to Linky from Enedis. The dedicated IS is based on a DMZ (demilitarized zone), i.e. a sub-network isolated from others and the Internet, closed to business players and accessible only by workstations that are not themselves disconnected from the intranet and the Internet (e.g., no emails). All of the provisions described systems are part of a dedicated PKI

(Public Key Infrastructure) to ensure the trust and durability of the system's security certificates. Finally, at the end of the chain, data that could be used by other Enedis information systems, is passed through controlled interfaces as encrypted flow towards the IS of other companies (RSA following the general ANSSI security reference).

Every household should have its own Linky meter. By 2021, Enedis has already deployed 35 million Linky meters, 770,000 data concentrators, 110,000 remotely controlled devices, replacing about 90% of old meters with a Linky meter. More information about the Linky meter rollout can be found in the documents "Le déploiement du compteur Linky" from the French Ministry of Ecology contains (Ministère de l'Écologie 2017) and "Report on the deployment of Linky smart power meters in the area" from 01 July 2019 (SMARTER TOGETHER 2019).

2.2 German Case

Protection of personal data from smart networks, in particular smart meters, is guaranteed at the legislative level by the Act on Electricity and Gas Supply (Gesetz über die Elektrizitäts- und Gasversorgung—Energiewirtschaftsgesetz, EnWG), which was published 7 July 2005. EnWG requires consumers consent before data collection and limits the data retention period.

The legal basis for smart metering rollout in Germany is described in "Act on the Digitization of the Energy Transition" (Gesetz zur Digitalisierung der Energiewende—GDEW). In February 2015, the Federal Ministry for Economic Affairs and Climate Action (Bundesministerium für Wirtschaft und Klimaschutz BMWK was BMWi) presented the key points for a regulatory package that was intended to promote the use of intelligent measurement systems security and cost-effectiveness. The measures required for the transition of the electricity supply to a decentralized system with bidirectional power and information flows includes:

- Avoidance of disproportionate costs for end consumers, producers as well as metering point and network operators in the conversion of 80% of end consumers to intelligent metering systems Smart Metering, which is required by EU Directives 2009/72/EG and 2009/73/EG.
- Minimal technical requirements to maximize the overall economic benefit from energy savings and load shifting.
- Provision of data protection and data security.

For the roll-out of the intelligent measuring systems (iMSys) in accordance with the law on the GDEW, a display of consumption and generation data that is protected against manipulation and conforms to calibration law is required for invoice verification purposes. End consumers should be able to carry out evidence-proof checks of bills from energy supply companies. In addition, the end consumer should be allowed to call up their current consumption values. More precisely this is described in the document Application Rule AR2418-6.

Moreover, the introduction of intelligent metering systems is tied to the compliance with a staggering price cap for annual costs to protect the end consumers from an extensive cost increase. Consumers have the option to choose an independent third-party metering operator if they are not satisfied with the solution offered by the Distribution System Operator (DSO), who in most cases is defined as the default metering operator if the consumer is not choosing a different operator.

On 2 September 2016, GDEW entered into force (Landis+Gyr 2020). The law also introduces specific and detailed requirements, both for the design of smart meter devices and for the transmission of data. GDEW's goal is to open the German energy market to digitization while ensuring a high standard regarding data protection and ICT security. GDEW's key feature is the introduction of smart meter gateways (SMGW). The gateways will provide each of the actors involved in supplying electricity with all of the information on generation and consumption that they need—from the grid operator or electricity supplier to the consumer. At the same time, smart meter gateways should provide the highest level of data privacy and data security.

The central element of GDEW is the Measuring Point Operating Act (Messstellenbetriebsgesetz, MsBG) which entered into force in September 2016 and regulates the metering points operation market and the equipment of the grid-bound energy supply with modern measuring devices and intelligent measuring systems. MsBG is divided into 4 parts:

- Part 1 regulates the scope and the terms used.
- Part 2 (§§ 3-48) regulates the operation of the measuring points, including the equipment of the measuring points with modern measuring devices and intelligent measuring systems, staggered over time and according to annual consumption.
- Part 3 (§§ 49-75) contains specific data protection regulations.
- Part 4 (§§ 76-77) determines the supervision by the Federal Network Agency (Bundesnetzagentur, BNetzA).

MsBG defines meter operation and measurement as a separate area of network operation that creates new market roles and has abolished electricity billing fees. It prescribes the comprehensive installation of modern measuring devices and intelligent measuring systems by the so-called "basically responsible metering point operator for modern measuring devices and intelligent measuring systems" (§2 No. 6 MsbG) by 2032. The norm addressees of the law are the German distribution system operators/supply grid operators (Versorgungsnetzbetreiber), considering that meter operation is dogmatically separated from network operation. As the basic metering point operators (gMSB), DSO's initially responsible for the rollout and administration of modern metering systems. As a result, the scope of tasks has expanded (previously metering device operation, measurement, and billing). Using a special public procurement procedure, they can however transfer this position to a third-party service provider.

MsBG defines extensive technical requirements for the technologies involved, particularly regarding the reliability and security of energy measurement and data transmission. Compliance with the new rules is controlled and supervised by both the Federal Office for Information Security (Bundesamt für Sicherheit in der Informationstechnik) and the Federal Network Agency (Bundesnetzagentur).

One more important document is the "intelligent metering systems rollout plan". This roadmap describes how gateways should be developed into a comprehensive digital communication platform for the energy transition ("Standardisierungsstrategie and zur sektorübergreifenden Digitalisierung nach dem Gesetz zur Digitalisierung der Energiewende Roadmap für die Weiterentwicklung der technischen BSI-Standards in Form von Schutzprofilen und Technischen Richtlinien"). The process comprises different rollout periods for different types of end consumers and plant operators, depending on the amount of energy they consume. For small consumers—whose annual consumption is less than 6000 kWh/a and/or feed-in systems < 7.5 kW peak, ca. 85% of the market or small producers of renewable energy or cogeneration—whose capacity is less than 7 kW, the installation of a smart meter remains optional because considered that the achieved energy savings would not return the invest of a SMGW installation. Customers with higher consumption and/or bigger renewable energy feed-in systems should install SMGW (ca. 15% of the market).

In Germany, the situation with DSO is different from France, because in the electricity sector exist 883 distribution network operators of varying sizes (Rullaud and Gruber 2020). The general rollout of intelligent metering systems began when BSI certified smart meter gateways (SMGW) from three manufacturers and issued a corresponding market declaration.

The Federal Office for Information Security (Bundesamt für Sicherheit in der Informationstechnik, abbreviated as BSI) is the German upper-level federal agency in charge of managing computer and communication security for the German government. BSI is responsible for the security of computer applications, critical infrastructure protection, cybersecurity, cryptography, counter eavesdropping, certification of security products, and the accreditation of security test laboratories. The BSI's scope of duties is defined by the German Federal Office for Information Security Act (BSI Act). The aim of the BSI is the preventive promotion of information and cybersecurity to enable and promote the secure use of information and communication technology in the state, economy, and society. As an example, the BSI develops practice-oriented minimum standards and target group-specific recommendations for action on IT and cybersecurity to support users in avoiding risks. BSI is also responsible for protecting the IT systems of the federal government. This involves defending against cyber-attacks and other technical threats to the IT systems and networks of the federal administration.

According to the requirements set by the "Metering Point Operating Act", the smart meter gateways must meet the security architecture defined by the BSI with regard to communication data protection and interoperability. These uniform technical and organizational specifications were written down in so-called protection profiles ("PP") and technical guidelines ("TR"), on which the BSI carries out the certification. The BSI has an extensive and well-structured documentation for the German smart metering infrastructure which describes at most every aspect of smart metering system work including requirements both for the architecture and the protocols used for deploying smart metering system. A full list of documentation is publicly available on the BSI website (https://www.bsi.bund.de/DE/Themen/Unternehmen-und-Organisationen/Standards-und-Zertifizierung/Smart-metering/smart-metering_node.html).

One of the most important BSI documents about SMGW is Common Criteria Protection Profile BSI-CC-PP-0073-2014. The Common Criteria (CC) for Information Technology Security Evaluation is an international standard (ISO/IEC 15408) for testing and evaluating the security characteristics of IT products. The protection profile describes possible threats to a smart meter gateway in its operational environment and defines the minimum requirements for appropriate security measures. The protection profile for the smart meter gateway focuses on the security performance to be met by a gateway and defines security requirements for the interfaces to the three networks (LMN, HAN, and WAN) that each gateway must provide.

Other important documents included the Technical Guideline BSI-TR-03109, which encapsulates requirements for functionality, interoperability, and security that the individual components in an intelligent metering system must meet. It specifies the security requirements and assumptions from protection profiles. The BSI-TR-03109 is divided into several parts:

- BSI-TR-03109-1 "Smart-Meter-Gateway"
- BSI-TR-03109-2 "Security module"
- BSI-TR-03109-3 "Cryptographic Specifications"
- BSI-TR-03109-4 "Public Key Infrastructure"
- BSI-TR-03109-5 "Other system units"
- BSI-TR-03109-6 "Smart Meter Gateway Administration".

After passing the Common Criteria Protection Profile certification a SMGW model gets a BSI-DSZ-CC-* type certificate. In case a SMGW passes the test specification "Testkonzept zu BSI TR-03109-TS-1" the SMGW model gets a BSI-K-TR-* type certificate. The TR-03109-1 certification includes not only hardware tests—but also software, and because of it, this certificate includes information about the software versions.

The market analysis was firstly published by the BSI on 31 January 2019. On 31 January 2020, the BSI updated the market analysis and published the administrative act to determine the technical possibility of installing intelligent metering systems. Currently, already 4 smart meter gateway models are certified by BSI rules (https://icube.ch/Security/security1.html).

With an urgent decision of March 4, 2021, the Higher Administrative Court of Münster (Oberverwaltungsgericht, OVG) in procedure 21 B 1162/20 has the suspensive effect of the main action pending before the Köln Administrative Court (Verwaltungsgericht, VG) against the general ruling of the Federal Office for Information Security (BSI) according to § 30 MsbG. The decision of the Higher Administrative Court in Münster was made in the form of provisional legal protection. The main decision by the Administrative Court in Köln is still pending. The BSI will therefore examine the OVG's reasons for the decision in detail and hopes to be able to comprehensively refute the OVG's concerns in the main proceedings (https://www.bsi.bund.de/DE/Service-Navi/Presse/Alle-Meldungen-News/Meldungen/Stellungnahme-OVG-Muenster-Smart-Meter_080321.html).

The Federal Network Agency (BNetzA) was not and is not involved in the proceedings before the administrative courts. However, as the supervisory authority for the MsbG, the Federal Network Agency must decide how to proceed with supervisory measures in the future. Due to the large number and complexity of the questions associated with the urgent decision, a final assessment is not yet possible. The BNetzA is currently also in close contact with the BMWK and the BSI on the consequences of the decision (https://www.bundesnetzagentur.de/DE/Beschlusskammern/BK08/BK8_09_MsbG/BK8_MsbG_Basepage.html).

BSI Smart Metering Architecture

The BSI smart metering architecture and protocols are strictly defined in BSI-TR-03109. A central element in the German smart metering system is SMGW (smart meter gateway). SMGW manages connections between 3 networks, stores, and checks consumption data (Fig. 2). The network formed by connecting electricity, gas, water, and heat meters to a smart meter getaway forms an LMN (Local Metrological Network). In HAN (Home Area Network) a SMGW communicates with the controllable energy consumers or energy producers (Controllable Local Systems, CLS, e.g. intelligent household appliances, heat, power, or photovoltaic systems, circuit breakers). The SMGW also provides data for the end consumer or the service technician in a HAN. Via a WAN (Wide Area Network) interface, external market participants and gateway administrators can access, configure and monitor the SMGW. The main functionality of the SMGW includes storing the measured consumption values received from the LMN, processing them according to configured rules, and sending the processed measured values to authorized market participants in the WAN (such as consumption and network status data). A SMGW fulfills the tasks of a firewall system and separates the connected networks from each other. As a decentralized storage of personal measured values, which are only sent to authorized parties under contractual regulations, the SMGW ensures data protection and data security for the end consumer.

A SMGW administrator is the most important actor in the German SMGW system. A SMGW administrator performs the following list of tasks:

- device management. Devices (meters, CLS, display units) must be registered by the SMGW administrator and assigned to an end user.
- client management. The SMGW administrator must create, edit, assign or delete assigned certificates or user ID/passwords.
- profile management. The SMGW administrator must have the opportunity to install and change meter, communication, and evaluation ACP (access control profiles) e.g. activate and delete tariffing and network status reporting.
- key/certificate management. The SMGW administrator must insert, activate, deactivate or delete keys and certificates for communication with meters, CLS, and external market participants.

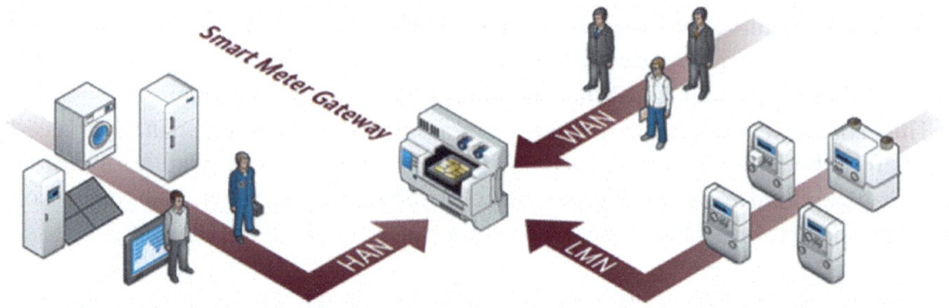

Fig. 2 German SMGW networks (https://www.bsi.bund.de/DE/Themen/Unternehmen-und-Organisationen/Standards-und-Zertifizierung/Smart-metering/Smart-Meter-Gateway/smart-meter-gateway_node.html)

- firmware update. The SMGW must allow the SMGW administrator to install, verify and activate new firmware in the SMGW. The SMGW admin must have mechanisms to verify the integrity before the activation of a SMGW can take place.
- configures, monitors, and controls an SMGW. The SMGW must allow the SMGW administrator to query the status of the SMGW and to read log entries from the system and calibration log. The GWA is not allowed to access privacy-relevant data, such as billing or meter data. The GWA can read a log with records of local attacks on SMGW.
- wake-up configuration. The SMGW must allow the SMGW administrator to configure the address of the wake-up service.
- can perform the GW operator role (delete CMS signature to pseudonymize data)

The German smart metering system uses two types of user data. Identifiable data used for billing purposes and pseudonymized data. Pseudonymized data is used by a grid operator to control the functioning of the energy network. By using pseudonymized data an operator cannot know who exactly consumed the reported amount of electricity because instead of a user name used an impersonal ID number and a grid operator gets this data integrally. That means that only the area in which the user is located is known. Since high-resolution energy consumption profiles can be used maliciously, the SMGW offers the possibility to calculate the total electricity cost locally at the SMGW and send it to the energy service provider once per month. By BSI requirements metering data must be stored directly in the SMGW. Data for up to 24 months can be downloaded at any time by the customer and shared with third parties. The supplier, who has the right to use the data, is obliged to delete all person-related metering data after the completion of his tasks.

Only SMGW administrators can install and change ACPs. An ACP defines which and how often data should be sent to external market participants and which cryptographic keys should be used for symmetric encryption and asymmetric data signing on the content layer. There are 3 types of profiles that determine how the SMGW collects and processes measurements from meters:

- Meter Profile: The Meter Profiles specify how a SMGW interacts with a Smart Meter. Among others, it configures the unique meter identifier, which protocol to use, which measurement registers to collect, information regarding encryption, and whether the communication is unidirectional or bidirectional. It also configures the time interval in which a SMGW must receive or request the measurements and update the internally saved latest meter reading.
- Evaluation Profile: The Evaluation Profiles specify which data need to be derived from the meter measurements. This is done by configuring a tariff use case and the respective parameters. Each examination profile contains a list of Communication Profiles determining the recipient of the generated data.
- Communication Profiles: There are different Communication Profiles for the HAN and the WAN. In both cases, the profiles define connection details that specify how to reach a communication partner based on uniform resource identifiers (URIs). For each of the three profile types, multiple instances may be configured, e.g., multiple Meter Profiles to collect measurements from multiple meters.

All secure keys and certificate materials (such as a private key for TLS authentication) must be stored in an SMGW hardware security module—HSM, which is a CC-certified (BSI-CC-PP-0077) subcomponent that is used for providing cryptographic operations. The smart meter gateway internally communicates with a security module via APDU commands (which are commonly used in smart cards such as SIM cards). More information about this can be found in BSI TR-03109-2. Full functional requirements for all three SMGW networks are presented in BSI TR-03109-1 paragraphs 2.3.1–2.3.4.

A SMGW communicates with meters only in the local metrological network—LMN. Locally connected meters have been made known to the SMGW by the SMGW administrator in the form of meter profiles. According to the TR, the LMN interface can be designed either as a short-range radio interface (wireless Mbus) or as a serial interface. Wireless connections have two variants—uni- and bi-directional communication. A uni-direction type of connection is required because a lot of smart meters could work only in uni-directional mode.

In cases of wired and wireless bi-directional communication connections are secured by the TLS protocol. In case of a uni-directional communication connections are secured by using AES-CBC+CMAC. The BSI TR-03116-3 (3.3.4) states that SMGW must be able to implement the role of the TLS server as well as the role of the TLS client to secure the communication links in the LMN. A SMGW must use HSM (hardware security module) for TLS handshakes and other cryptographic operations. For mutual authentication between a SMGW and meters in the LMN, LMN certificates which are X.509 self-signed certificates must be used.

The data transmitted by the connected meters in the LMN can be consumption values as well as information on energy quantities fed into the grid (e.g., in case of photovoltaic systems, combined heat, and power plants). Also, further parameters relevant to grid operation such as grid voltage, frequency, and phase angle, which may be provided by a meter, can be recorded by the SMGW. The following processing steps are performed by the SMGW at the LMN interface:

1. The SMGW receives or retrieves the measured values of the locally connected meters at regular intervals. The SMGW receives the measured values in an encrypted form and securely integrates them.
2. After successful decryption and integrity check of the measured values, the SMGW provides them with a timestamp provided by the system clock of the SMGW and stores them in measured value lists.
3. The SMGW determines derived value from certain measured values using a set of rules and sends these processed values to authorized external market participants.

For reasons of data protection, BSI has precisely defined how the software must process the measured values according to the respective evaluation profiles in the gateway. There are various tariff use cases (TAF) for tariffing, balancing, and network status data collection, which must be implemented as a minimum requirement of SMGW through regulations.

According to the BSI's standardization strategy, the conformity of Generation 1 smart meter gateways with calibration law is evaluated based on application rule 50.8 defined by the National Metrology Institute of Germany (Physikalisch-Technische Bundesanstalt or in short PTB). Currently, there exist 14 tariff use cases, but Generation 1 smart meter gateways support only TAF 1, 2, 6, and 7.

After collecting data from a smart meter the SMGW should check it, and send it to external market participants (EMP or EMT in German) in Wide Area Network (WAN). The WAN interface is designed as an IP interface. Usually, the SMGW uses a cellular network to connect to external market participants (Fig. 3). There also exist solutions that use Ethernet as a WAN port, but in practice, they are less common.

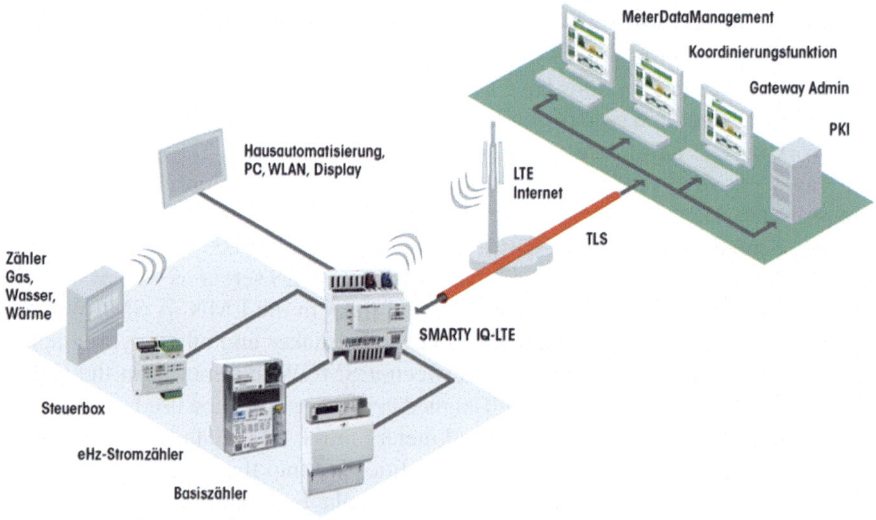

Fig. 3 German smart metering architecture (Siconia® SMARTY IQ-LTE | Sagemcom)

The Smart Meter Gateway is integrated into a public key infrastructure via the WAN interface and all communication is encrypted with TLS. A WAN connection must always be established as a mutually authenticated TLS channel based on certificates. The used certificates must be issued by the Smart Metering Public Key Infrastructure (BSI 2017). In cryptology, a public key infrastructure (PKI) is a system that can issue, distribute and check digital certificates.

All WAN use cases and communication scenarios are described in BSI TR-03109-1 chapters 3.2.2–3.2.3. For the communication with participants in the WAN, the SMGW must always implement the role of the TLS client and the remote terminal the role of the TLS server—which means that all communication connections must always originate from the Smart Meter Gateway. WAN connections can be established by the gateway at specified times (e.g., by a timer trigger), or by spontaneous events. The ability to react to spontaneous events requires the use of a SMGW wake-up service (WAF 7)—which is the only way to establish a connection with a SMGW from outside. Only SMGW administrators can get access to this service using valid cryptographically signed messages because a SMGW should not respond to any other connection attempts. A more complete description of the wake-up service could be found in BSI TR-03109-1 chapter 3.2.5.

Data in WAN SMGW communications are not always transmitted via a direct transport channel between a sender and an end recipient, but sometimes via third parties (e.g., the gateway administrator). This type of data exchange between communication partners in the WAN takes place within a TLS channel based on messages encrypted and signed at the content level for the end recipient via CMS (Cryptographic Message Syntax). CMS is a special set of data types each of which can be represented as a box performing a certain action, for example—encryption. CMS types Signed-data Content-Type and Authenticated-Enveloped-Data Type are used to sign and encrypt data. More information can be found in the document "Technische Richtlinie BSI TR-03109-1 Anlage I: CMS-Datenformat für die Inhaltsdatenverschlüsselung und -signatur". CMS is used for data pseudonymization. A SMGW replaces the client identification number with a pseudonym, encrypts, and signs the data at the content level. Only the energy service provider has the key to decrypt this data. In the case of pseudonymized transmission of measured values (network status data), the SMGW must replace the canonical device ID of the SMGW and the canonical device ID of the logical device from which the measured values originate with a pseudonym. The pseudonymization of network status data during transmission from the SMGW to an external market participant must be ensured by the following steps:

1. The SMGW removes the unique canonical device ID from measured values that are to be transmitted pseudonymized following an evaluation profile and replaces it with a pseudonym stored in the evaluation profile.
2. Data prepared in this way are then encrypted by the SMGW for the recipient (EMT), signed, and transmitted to the SMGW administrator.
3. SMGW administrator checks the signature of the SMGW, thus verifies the authenticity of the received data and forwards it to the recipient after removing the SMGW signature.
4. The recipient decrypts the message.

Tracing the final consumer based on the signature of the sending SMGW is much more difficult for the recipient (in the example: EMT-A) because the SMGW signature has been removed by the SMGW administrator. Tracing the final consumer via a canonical device ID is considerably more difficult for the recipient since the data contains a pseudonym instead of the canonical device ID.

For metering and gateway administration used COSEM IEC 62056-6-2 with OBIS (Object Identification System). COSEM (Companion Specification for Energy Metering)—is a specification that reflects an interface model of the metering devices to represent their functionality. The interface model uses an object-oriented approach.

Home area networks (HAN) are networks that ensure the networking of computers and their peripheral devices in smart homes and home offices. For authorized end-users, a SMGW provides the possibility to retrieve information stored in the SMGW through the end-user interface in HAN. This scenario is implemented by HAF1: End-user data provision. This data can only be accessed in read-only mode and only after successful authentication. The reading and visualization of the data at this interface, requires a dedicated cryptographically secured display, local PC, or another (CLS) device that can process the cryptographically secured data stream. Other SMGW functions in the HAN interface include:

- HAF2: Service technician data provision.
- HAF3: Transparent communication channel between CLS and active EMT.
- HAF4: Establishing GWA communication by a service technician.
- HAF5: Triggering of self-test functions by a service technician.

There are 5 HKS (HAN communication scenarios) for SMGW in HAN (Home Area Network):

- HKS1: Bidirectional communication in HAN with authentication using HAN certificates.
- HKS2: Bidirectional communication in the HAN with authentication using a unique identifier and Password.
- HKS3: Transparent communication channel initiated by CLS.
- HKS4: Transparent communication channel initiated by active EMT.
- HKS5: Transparent communication channel initiated by SMGW.

As shown, a connection can be initialized by an EMT, the SMGW, or the CLS initiative, but it previously requires a registration by the SMGW administrator. HKS1 can be used by the end-user to retrieve consumption data or by a service technician to obtain technical data. In HKS1 the SMGW implements the role of the TLS server and the participant implements the role of the TLS client. When establishing the TLS connection between the HAN participant and the SMGW, client–server authentication is performed in the TLS handshake using the GW_HAN_TLS_CRT and CON_HAN_TLS_CRT certificates and their associated keys. The certificate CON_HAN_TLS_CRT is uniquely assigned to an end-user or service technician known to the SMGW.

In HKS2 the SMGW implements the TLS server role and the end-user implements the TLS client role. A service technician must not use the communication scenario

HKS2 and should use a special certificate for authorization. As mentioned before, the SMGW also supports the remote control of CLS (controllable local systems) by EMP (external market participants). For this purpose, the SMGW provides HKS3, HKS4, and HKS5 using communication channels secured by TLS within which an EMP can communicate with a CLS using any protocol that works over TLS.

To be able to use HKS3 CLS should support the draft RFC "Secure Sockets Layer for SOCKS Version 5". By default, SOCKS5 does not encrypt traffic and for security purposes, this draft RFC requires establishing a TLS channel before the main connection establishing. In the case of HKS3 CLS initiates the communication by connecting to a SMGW SOCKS5 port. The SMGW answers by indicating the support of the X'86' authentication method only. The CLS in the role of TLS client initiates the TLS connection. The SOCKS5 protocol in this case is required to redirect data from the SMGW specified in Socks5 message EMP. Before establishing this connection, the SMGW should check if the SMGW administrator allows the connection of this CLS to the requested EMP by checking the communication profiles. Because direct connections from the WAN to SMGWs are forbidden, in HKS4 the EMP must first set up the backend system TLS server port and send a connection request to the SMGW administrator with the necessary CLS information (proxy profile ID and connection profile ID). Then SMGW administrator wake-up SMGW and use these IDs to establish a connection to an EMP. In this case CLS should be able to implement a TLS server role. The SMGW implements the role of the TLS client both in the HAN and the WAN. In HKS5 a connection is initialized by a SMGW trigger which could be some timer or event. As in HKS4, in HKS5 the SMGW implements a TLS client role in HAN and WAN and the CLS implements a TLS server role.

In all 3 scenarios of communication with an EMP the SMGW acts like a proxy server for controllable local systems (CLS) connected to the HAN. It means that TLS-protected communication channels in the direction of the CLS and the external market participant are terminated in the SMGW and the SMGW takes over the transparent forwarding of the received data. The current version of the TR requires the implementation of HKS1, HKS2, and HKS3 only. The implementation of HKS4 and HKS5 is optional. HAN use cases, communication scenarios, and profiles are described more precisely in BSI TR-03109-1 chapters 3.4.2.1–3.4.2.3, 3.4.3.1–3.4.3.5, 3.4.6.2.

Also, the HAN interface is used by the service technician to view configuration profiles and the system log that supports error diagnosing. The SMGW logs all actions in three different log levels, the system log, the end-user log, and the calibration log. Every important event (e.g., error messages, failure of the WAN connection, security-relevant events, activities of the SMGW administrator, etc.) in the SMGW is logged in the system log. This log can only be viewed by the authorized SMGW administrator and the authorized service technician on site. The information is used to identify the current status of the SMGW and to identify possible sources of errors or malfunctions.

All SMGW transactions (e.g., the sending of measured values and activities of the SMGW administrator) are recorded in a final consumer log. An authenticated and authorized end user can call up the relevant information from the SMGW via the logical HAN interface for display units and thus keep track of who has received which data, when or whether user-related data (e.g., profiles) have been changed, added, or removed. To maintain the confidentiality and integrity of the personal log data, an SMGW administrator is not allowed access to the end-user log.

The calibration log stores events relevant to calibration (e.g., detected falsifications of measurements or failed time synchronization). Besides, changes to parameters that are relevant to calibration technology are registered here (e.g., setting of the device clock). This log can only be viewed by the authorized SMGW administrator and verified by the SMGW administrator authorities if it is required.

Authentication for getting logs is also important because the SMGW must be able to record and save the measured values from meters of various final consumers (for example in multi-family houses). For this purpose, the SMGW has implemented mechanisms to support multi-client capability and the associated authentication requirements from GW_PP chapter 1.4.6.6.

2.3 Swiss Case

Due to the heterogeneity of network operators in Switzerland, implementation of the standards varies depending on the size of the network operator. Currently, there is a parallelism between federal and cantonal law in the area of data protection in Switzerland. In particular, federal data protection legislation, which also regulates data security, does not contain sector-specific rules, but general rules whose application in a specific case may leave considerable room for interpretation. In addition to the Federal Data Protection Act, which applies to private individuals and federal authorities, the cantons also have their data protection laws that apply to cantonal authorities. Since the vast majority of network operators are part of the cantonal administration in the broader sense (cantonal utilities), such network operators are affected by cantonal laws. This parallelism of federal and cantonal law leads to legal uncertainty, particularly in the operation of smart metering systems. This legal fragmentation, particularly concerning granularity and the use of load profile data, can impair the benefits of smart metering systems (UVEK 2015).

The core document about Swiss smart metering was provided by Federal Council (Schweizerische Bundesrat) and named Stromversorgungsverordnung (StromVV, Electricity Supply Ordinance). Articles from Art. 8a to Art. 8d contain information about (Schweizerische Bundesrat 2008):

(a) Smart metering systems
(b) Data security check
(c) Intelligent control and regulation systems for network operation
(d) Handling of data from intelligent measurement, control, and regulation systems.

From the technical point of view, these articles contain very mild limitations, such as:

- The smart meter should have a physical display on the smart meter case.
- The smart meter must have local data storage. This storage is necessary to store information in case there is no connection to the server, SMGW, or data concentrator. Also, this storage should be enough to store load profiles with a 15 min period for at least sixty days.
- The smart meter should have interfaces for bidirectional communication with a data processing system and enable the possibility to get measured values at the moment of their acquisition as well as the load profiles to the end-user, producer, or storage operator.
- Manipulations and other external influences on the smart meter should be detected, recorded, and reported.
- Other digital measuring equipment, as well as intelligent control and regulation systems of the network operator can be integrated with a smart meter.

Also, this document contains customer data management restrictions. The personal data and personality profiles should be destroyed after 12 months unless they are relevant to billing or have been anonymized. The network operator shall retrieve the data from intelligent metering systems a maximum of once a day unless network operation requires more frequent retrieval.

Network operators may process data from the use of measurement and control systems without the consent of the data subject for the following purposes:

(a) personality profiles and personal data in pseudonymized form, including load profiles of 15 min or more: for measurement, control, and regulation, for the use of tariff systems and secure, efficient, and effective network operation, network balancing, and network planning;
(b) personality profiles and personal data in non-pseudonymized form, including load profile values of 15 min and more: for the billing of energy supply, grid usage charges, and remuneration for the use of control and regulation systems.

Network operators may pass on the data from the use of measurement systems to the following persons without the consent of the person concerned:

(a) personality profiles and personal data in pseudonymized or suitably aggregated form: to the parties involved by Article 8 paragraph 3;
(b) the information needed to decipher the pseudonyms: the energy supplier of the final consumer concerned.

Requirements from StromVV were checked and supplemented by DETEC and VSE at the next steps. DETEC—Federal Department of Environment, Transport, Energy and Communications (Generalsekretariat des UVEK). DETEC is one of the seven departments of the Swiss federal government, headed by a member of the Swiss Federal Council. This department is responsible for issues related to environmental policy, management and development of transport, management, and monitoring of energy sources (electricity, gas, oil, etc.), and mass media (including television).

VSE (Verband Schweizerischer Elektrizitätsunternehmen) is the main organization, educational institution, and political mouthpiece for the Swiss electricity industry. Its members ensure over 90% of Switzerland's electricity supply. VSE creates industry recommendations, manuals and also develops future scenarios, basics, and positions in the electricity field. The most important VSE documents about smart metering are "Smart Metering System Data Security Guidelines" (VSE 2018) and "Intelligent measuring systems. The use of intelligent measuring systems in Switzerland" (VSE 2019). VSE recommendations are more concrete compared with StromVV. As an example:

3.4.1 Firmware—operating system and applications in all major components are subject to version control; are installed upon delivery and secured against unauthorized commissioning. The integrity of this data is a prerequisite for proper and trustworthy operation. Firmware update—operating system and applications in the main components are subject to version control; are installed and put into operation in the main components according to the version control by authorized users in the role of administrator. The integrity of this data is a prerequisite for proper and trustworthy operation.

6.2.5 Customer interface. The customer interface should be located on the meter. The communication should be encrypted according to the data security check. Protocol and interface are designed manufacturer-specific (e.g. DLMS via RS485). Visualization systems that can be used also depend on the manufacturer and it is not standardized. To be able to offer customer support in the selection of third-party products, it is recommended to consult the supplier of the iMS for possible references.

Swiss Smart Metering Architecture

Possible smart metering architectures are considered in "Smart Metering System Data Security Guidelines" (VSE 2018) in article 2.5 and "Intelligent measuring systems" (VSE 2019) in 6.3.2.

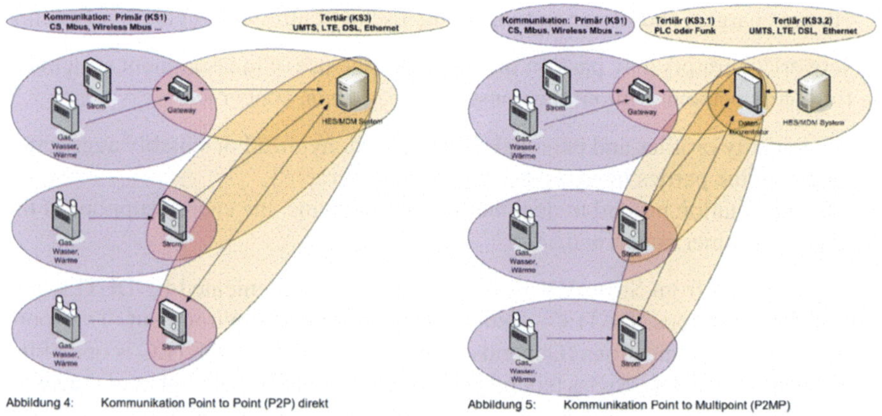

Fig. 4 Swiss smart metering architecture (VSE 2019)

As we could see from Fig. 4, using of smart meter gateway or data concentrator is optional. Gas and water meters can transmit data through an electricity meter. Possible options for the in-home architecture from [Richtlinien für die Datensicherheit von intelligenten Messsystemen, Anhang 1] includes:

1. Smart meter devices connected in parallel (without data concentrator)
2. Smart meter device with LMN
3. Smart meter device with cascaded LMN
4. Gateway with LMN
5. Gateway with cascaded LMN.

In the document "Basics of designing an introduction of intelligent measuring systems at the end consumer in Switzerland Minimum technical requirements and introduction modalities" (UVEK 2014) the Swiss Federal Office of Energy evaluated the appropriateness of different smart metering approaches by defining minimum national requirements and comparing them with implementations in other countries, resulting in a large overlap between the Swiss requirements and the German approach.

To be officially used in Switzerland smart metering system should obtain a certificate from the METAS-Cert (Eidgenössisches Institut für Metrologie Konformitätsbewertungsstelle; https://www.metas.ch/ds) In accordance with the provisions of Electricity Supply Ordinance Article 8, it must be proven for all elements used that they meet the data security requirements resulting from the Swiss protection requirements analysis. Certification proof methodology could be found in the document "Prüfmethodologie zur Durchführung der Datensicherheitsprüfung für Smart Metering Komponenten in der Schweiz" (swissmig 2019). This document contains concrete security testing requirements divided by one section per type of smart metering system element:

- Section 5.1—General requirements—apply to all main components (Hauptkomponenten, HK) of an intelligent measuring system (iMS) include four sheets numbered with Roman numerals in this representation
- Section 5.2—Requirements for the intelligent measuring device (iMG)
- Section 5.3—Requirements for the gateway (GW) as a communication system (KS)
- Section 5.4—Requirements for the data concentrator (DC) as a communication system (KS)
- Section 5.5—Requirements for the head end system (HES)
- Section 5.6—Requirements for key management (KM).

For example, requirements for intelligent measuring device interface KS2 from chapter 5.2.2.3 include:

(a) At least the user role prosumer is available to access this interface according to the corresponding access rights.
(b) Authentication takes place at least via user name and password.
(c) The interface allows the prosumer role to have read-only access to the count data intended for visualization.

(d) No connection to other interfaces of the iMG is possible via the interface.
(e) The interface is hardened against attacks such as denial of service, replay, buffer overflow, etc.
(f) A failure of the interface does not affect the metrological part or the other interfaces.
(g) Unauthorized access attempts and other disruptions trigger an alarm to the MDM system and these events are included in the log data.

As we can see, these requirements are close to the requirements from BSI protection profile BSI-CC-PP-0073-2014, but compared to German requirements from TR-03109-1 still more abstract because of the lack of predefined communication scenarios, running services, etc. Because these important elements are not defined—it is complicated to produce more concrete requirements than requirements that are currently defined in the "Test methodology" document.

In differ from the French situation with Enedis, the Swiss Federal Electricity Commission counts over 630 active DSOs for the 8.6 million Switzerland population in the year 2020 (Rullaud and Gruber 2020). In this document, we will compare the smart metering systems of two major manufacturers—Landis+Gyr and Siemens—that are used in Switzerland due to the large number of DSOs and relatively abstract supervisory requirements.

Landis+Gyr Smart Metering Infrastructure

The following sections considers a Landis+Gyr E450 smart meter and the DC450 data concentrator configuration. Landis+Gyr was the first supplier to receive data security certification from the Federal Institute for Metrology (METAS) for a G3-PLC data concentrator on 16th March 2021. A few weeks later Landis+Gyr's E450 smart meter and Head-End-System HES were equally certified resulting in Landis+Gyr now covering the complete chain of a smart metering system in an end-to-end certified solution. Landis+Gyr entrusted CCLab in 2019 to evaluate three components by the Swiss Smart Metering protection requirements (https://www.cclab.com/news/landis-gyr-metas-certificate):

- E450 is a residential advanced meter with an integrated PLC modem,
- DC450 is a new generation intelligent data concentrator for a large-scale meter reading and controlling applications and
- the HES which is an interoperable head end system that provides a communication and data collection layer between the smart meter infrastructure and the utility.

This architecture looks very similar to French Linky system architecture. It does not have smart meter gateways (smart meter already plays the role of meter and communicating unit in one box), has a data concentrator, uses PLC at the physical layer, and has the ability to connect household appliances to manage power (Fig. 5).

To secure data Landis+Gyr E450 smart meters use DLMS/COSEM security suite 0 with High security level. DLMS security is always used independently of the lower communications and security layers, which means that data concentrators just resend data without decrypting it. To prevent interception or falsification of messages

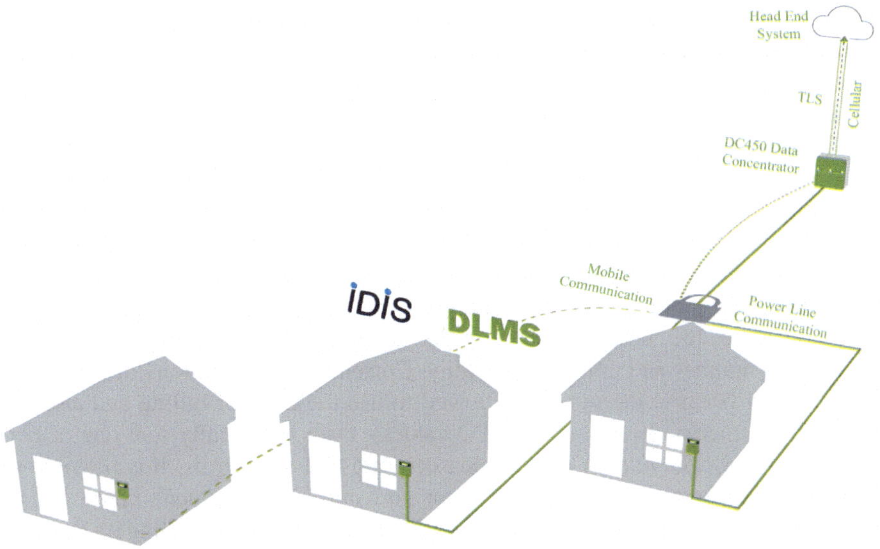

Fig. 5 Landis+Gyr smart metering solution architecture (https://www.landisgyr.de/product/landis gyr-concentrateur-dc450-2/)

and feed unauthorized commands between the HES (head-end system, in German SMI terms, is one of EMP), data concentrator, and smart meters used DLMS or TLS encrypted and authenticated data (Landis+Gyr 2020). Communication security based on the IDIS DLMS/COSEM standard which includes:

- Symmetric encryption and keys for meter communication with previously exchanged keys are used to enable good communication performance with embedded devices.
- 4-stage client-server authentication with GMAC.
- Recognized and proven AES encryption, 128-bit key.

In case to implement secure communication key management:

- Secure keys are created and stored in an encrypted format on devices at the time of manufacturing.
- The key material is encryptedly sent to the customer's HES system.
- The keys are stored in Landis+Gyr's HES Secure Key Manager.
- Once communication is established, the keys can be exchanged regularly or on-demand.

Most of the Landis+Gyr smart meters have from at least one to several control outputs to switch loads. There is also the demand for supervision in the meter which means if the demand/used current reaches a pre-defined limit, loads can be switched off/on (depending on the parametrization and load connected). The control output can also be controlled from the system side, reacting to information not available at the meter point. User access to functions and data is securely controlled based on the user's role M-Bus (authentication, authorization). Future security upgrades can be securely distributed, stored, and activated. Connection via local interfaces such as CII (Customer Information Interface), M-Bus, or wireless M-Bus, and optical interfaces can also be encrypted if required.

Only the utility has access to the smart meter data. Other parties can get their data from a central data hub in the cloud or do not get smart meter data at all (depending on concrete place laws). Landis+Gyr smart meters currently do not support multiple users of one device—every consumer needs their own smart meter to measure consumption. The data coming from a smart meter is only related to the serial number of the meter. Only the utility can make the connection between the meter serial number and consumer identity. Currently, data pseudonymization was not necessary because the utility is allowed to use the data for billing and also for grid stability checking (e.g., 15 min PQ values). Data is typically sent one or two times per day. Locally smart meters measure data at 1-s intervals. It is possible to push these data every 5 s over a consumer port into the consumer's home. Smart meters usually store data in meters from 10 days up to 1 year, depending on the kind of data. Data can be sent every 15 min or more often. The most significant limitation is the used communication technology and the cost for the utility to data transfer, but modern smart meters are flexible in this regard and can be parametrized towards the use case.

In order to get an insight in Landis+Gyrs security, the Gridstream (Landis+Gyr 2020, 2014) solution which is usually integrated in the above described devices is considered in the following. Gridstream is a powerful energy management service based on two-way data communication, which enables a wide range of applications, such as remote meter reading, customer relationship management, and demand-side management. Gridstream provides functions to assist utilities with load control, reporting power outages, and monitoring power quality. The data flowing through the Gridstream system is exposed to various risks, such as intrusions in the field network, the data center, or even at a system level. The security architecture for Gridstream ensures system and network availability, while at the same time meeting critical security objectives, such as confidentiality, integrity, and authentication of data. The key elements of the Gridstream security approach include:

- PKI environment for managing security certificates. This infrastructure forms the basis for the management of secure communication, also on the LAN and WAN level.
- HSM module—The HSM serves as the root of trust where the utility ECC private key is vaulted. The private key is used to generate digital signatures to downstream commands sent by the HES. The HSM also features FIPS 140-2 and Common Criteria Level 4 certifications, providing strong protection for one of the critical elements in the advanced security architecture.
- Role-Based Access Control—The HES enforces access controls through a Role-Based Access Control (RBAC) functionality. RBAC allows the security administrator within a utility to manage user credentials and privileges assignment. In

this way, the utility can manage which employees have access to commands and features related to the devices in the network.

- Meter Tamper Alarms—Landis+Gyr meters offer tamper alarms such as reverse energy flow, tilt/tamper, and outage notifications in case a meter is removed from the socket. The alarms are transmitted to the HES upon occurrence and logged by the head end, displayed on the GUI, and optionally emailed to appropriate users. In the case of the network gateway or network bridge cover removal, an alarm will be immediately sent to the head end, displayed on the GUI, and optionally emailed to the appropriate users.
- Firmware Integrity—All firmware images released by Landis+Gyr are digitally signed utilizing the Landis+Gyr asymmetric private key (ECC). Each endpoint within the network will validate the digital signature using the Landis+Gyr public key. If the key doesn't match the Landis+Gyr signature, the device will not upgrade that firmware version. This is a strong prevention mechanism to avoid the injection of rogue code into the network.
- From the factory, all Landis+Gyr IDIS devices are produced with a security key set in an ISO27001 certified environment. These are stored in encrypted form in the meter and stored in a hardware security module at Landis+Gyr. Later, as part of the activation of the communication security process, these keys are transmitted to the certified customer as a file using a secure procedure.

The servers in the head-end system platform are all protected using hierarchical access control mechanisms based on access rights (roles). The internal network of AIM system platform servers is typically protected from external access using firewalls and multiple levels of network access control mechanisms. All users of external connections should be authenticated inside a secure data connection protocol. This is especially important for protection against signals from an unauthorized meter or a computer that emulates a meter. The authenticity of a meter that is sending data is ensured using the DLMS-COSEM high level security authentication protocol. The DLMS-COSEM protocol ensures secure access to the electricity meter's data. Data access security is based on assigning different access rights (Fig. 6).

Remote reading of metering values from a point-to-point metering device (e.g., a device that communicates with the head end system using a mobile network) by the data collection system is typically protected using a virtual private network (VPN) tunnel between the internal Gridstream network and security enforced by the communication service provider and network-provider. In the case of a local area network (LAN) connected meter, a VPN with a firewall may be enforced between the LAN meter and the AIM internal network.

Field tools and data collection system applications—for operating the AIM site manager application, the site manager typically runs on a server in a de-militarized zone (DMZ) and is connected via a VPN tunnel to the telecom operator network with an optional HTTPS connection between the field tool device application and the site manager server. For AIM dashboard, it is advisable to run the dashboard server in the DMZ, and those HTTPS connections are enforced between the user terminal and

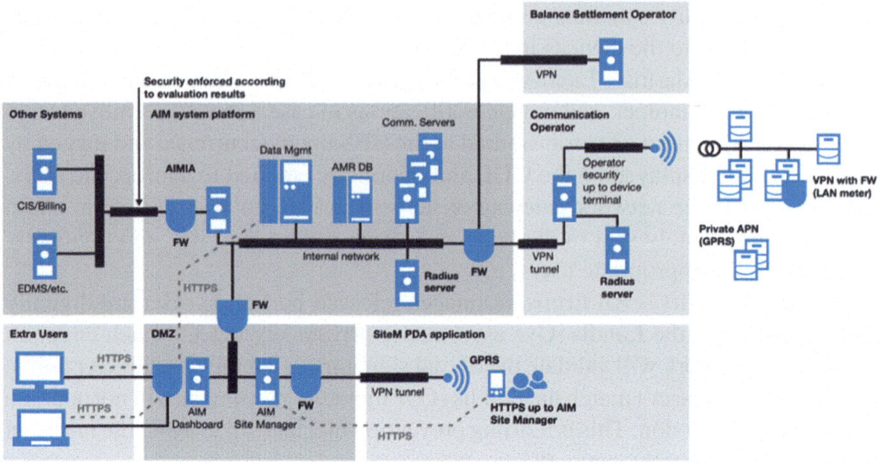

Fig. 6 Example Gridstream AIM infrastructure using sub-networks and zones (Landis+Gyr 2014)

the dashboard server. The data transfer between the AIM Dashboard server and the AIM system platform server may also use the HTTPS protocol.

Gridstream includes automatic repeat functions at different system levels to enable smooth recovery and to solve any possible communication problems. The communication protocols used have automated control and correction functions. Communication between the measurement, display device, and data collection system is typically implemented via a cellular network or power line communication (PLC). Message security over the mobile telephony network is typically provided by the wide-area networking VPN functionality of the network provider, in conjunction with the use of HTTPS on the transport layer. Providing message security over power line communications has been a subject of work in the IDIS consortium and DLMS-COSEM standardization. Confidentiality, integrity, and authentication are provided by using AES encryption and key distribution services throughout the measurement, switching, and display system itself, and at the interfaces to the data collection system. Communication within a meter is done by using an internal bus. For instance, meter values are read directly from the meter's measurement integrated circuits by using the meter's internal bus. The internal bus is only accessible by breaking the meter cover seal, which sends a system alarm to the data collection system, making such security attacks easy to detect and locate.

In addition to securing communication, Gridstream ensures that all data is safely stored and transferred. Each metering value coming from the meter has a status indication with the assured time of measurement, to show whether the user can trust the value or not. It shows, for example, if there were power cuts or if the device's time was adjusted during the measuring period. This information helps the user to locate the details of the device in question, if necessary. The Gridstream logging and auditing functions ensure that all modifications to the system data are carefully traced in the system.

Siemens Smart Metering Infrastructure

The Siemens architecture with IM150 or IM350 Smart Meters and smart meter gateway SGW1050 is very close to the Landis+Gyr architecture, but instead of "data concentrator" it uses a "gateway". Whereas traditional "data concentrators" include intermediate storage of the meter data, due to security reasons the Siemens SGW1050 gateway transparently and securely tunnels the meter data on their way to the HES. There is no decryption/re-encryption or insight into meter data on the gateway. The smart meter gateway SGW1050 can work not only with Siemens smart meters and can handle up to 2000 smart meters.

A smart meter measures electricity and communicates with the central system. For this reason, a communication module is used which can be either integrated or replaceable that allows the use of different communication technologies like G3-PLC, mobile, or some other. Since both parts have to go through separate certification processes, the firmware for both parts is split which allows updating communication relevant device logic without touching the measurement-related part of the device firmware.

Each smart meter can manage up to 4 m connected via wired or wireless M-BUS (OMS and IDIS Standard). The Siemens IM350 Smart Meter is using a wired MBUS interface to allow suppliers of secondary meters to connect their own wireless adapters to communicate with their meters. This concept works particularly well in regions where the smart meter is housed in a closed outdoor box with limited wireless connectivity. The smart meter collects consumption data, stores it internally, and sends it to the HES via a communication channel. On the HES level, those 4 m are independent and the orchestration is maintained. After transferring the data to an MDMS they will be interpreted as different channels (electricity, gas, water, heat, district heat) coming from the same SDP (service delivery point). A SDP describes a specific connection point in the field.

The master data system (billing system) can choose a measurement profile (data that the meters have to collect and should be stored in HES/MDMS for information purposes) and a billing profile (data that shall be sent further to the billing system for accounting). Both types of data are transmitted in pseudonymized form. There are register-based data (daily) profiles, interval-based data profiles (on 15 min schedule), or both. The kind of selected profiles depends on the purpose of the meter. Residential meters need to be billed on a monthly period but can collect 15 min values as well (to receive an in-depth network load picture per 15 min). Small industrial and industrial meters have to follow a specific schedule and need to be billed daily. After running a plausibility check and if necessary, a substitute value creation process, this data will be exported to an Energy Date Management system like BELVIS. Whichever profile has been set, all meters will send their data daily to the central system.

If a meter couldn't be reached on a specific day, the HES will request all missed data as soon as the device is reachable again. Due to network interferences and interferers, it's quite common by using PLC technology to receive on one daily readout roughly 80% of the installed smart meters. After 2–3 days the central system receives missed data as well until it has collected around 95% of the total data for a specific day. The smart meter stores its data for a minimum period of 60 days locally. The head-end system tries to read out each meter per day and can collect missed data as long as the meter has been stored locally. If a meter is not communicative for a longer period, a service technician has to care about this specific device. Instantaneous energy consumption is always counted and never deleted (counter reading). Register-based or interval data will be stored out of the instantaneous values according to their defined schedule (daily at midnight and interval per each 15 min). This allows all data to be collected in the aftermath, even when the meter was powered off in the meantime.

Electrical data is stored in the OBIS Code registers form. Such a register refers to a specific value and its metadata. All these registers can be stored on four different tariffs. Most of the utilities in Switzerland use just two of them (High- and Low Tariff). In the future, there will be no tariff-based storage on the smart meters anymore because the MDMS can deviate it directly into the central system (necessary for dynamic tariffing/pricing). Siemens IM350 can store 64 different energy-related registers. Most of the customers are using profiles containing 4–17 registers.

Since the gateway only transmits information but does not decrypt it the security features of the DLMS/COSEM protocol are used in conjunction with TLS. The customer interface uses IDIS CII standard encryption for transmitting information securely using AES-128 encryption. To secure communication channels RSA TLS with at least 4096-bit key lengths is used. Asynchronous encryption is ensured by certificates issued for a specific key usage ran under a dedicated Public-Key Infrastructure. To secure meter encrypted data 48 role-based meter keys are used. A role-based access concept ensured by individual keys and certificates is implemented for each device and application.

Siemens makes use of HLS with DLMS/COSEM Security Suite 0 and 1. NLS is not used at all. The DLMS/COSEM Standard belongs to this kind of standard which leaves open quite a big amount of interpretation. For this reason, Smart Meter suppliers in Switzerland implemented the IDIS Standard (Package 1-3) which is, in fact, a tight definition of how the DLMS/COSEM standard has to be implemented. This allows smart meters of different vendors to act interoperable.

Besides, Siemens is using dedicated appliances such as a hardware security module. This module is responsible for the whole meter data decryption and is the only system on which clear text meter keys are present. Meter keys never leave the tamper-resistant foil-coated CPU or its cache. This ensures a physical protection layer to the smart meter data security.

3 Comparison with Reference Architecture

As a base for architectures and used protocols comparison a reference architecture for smart metering communications of the Smart Metering Coordination Group was defined in the framework of the smart grids M/490 mandate and that for smart meters M/441 mandate (CEN/CLC/ETSI/TR 50572, 2011) which includes (Fig. 7):

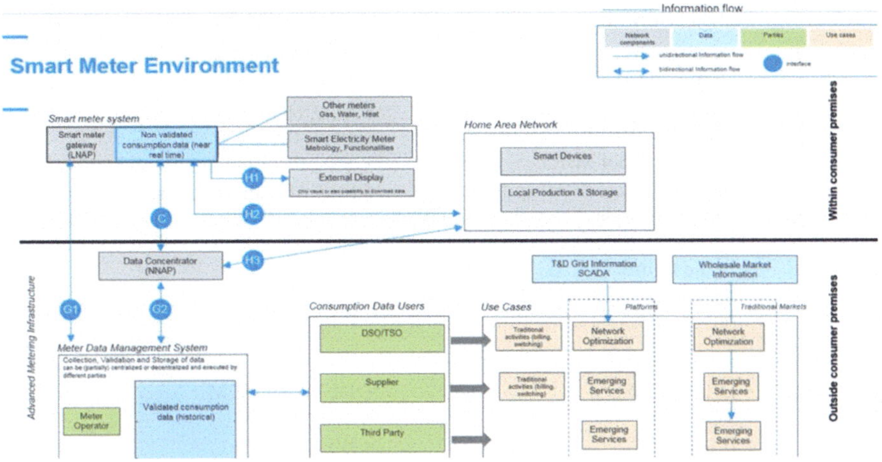

Fig. 7 Reference architecture with protocol labels (European Commission DG Energy 2019)

- Interface **H1** connects the smart meter system to an external display, via one-way communication. The external display is not uniquely designed. For instance, information may be provided only visually, or be available for download. In BSI terms interface **H1** is used in HAF1 for the end-user data provision in the Home Area Network (HAN).
- Interface **H2** connects the smart meter system with the Home Area Network (HAN). The HAN interconnects smart home devices for energy management purposes. Interface **H2** provides two-way communication, i.e. the HAN can send information on individual devices back to the smart meter system. In BSI terms interface **H2** is required to communicate with Controllable Local Systems (CLS).
- Data from the smart meter is shared externally with the meter data management system (a central communication system). This system communicates with meters either directly through the Wide Area Network (WAN) and enabled by interface **G1**, or via a data concentrator where information from several meters in a neighborhood is concentrated (Neighbourhood Network Access Points, NNAP) and enabled by interface **G2**.
- Interface **C** is used to connect LNAP (Local Network Access Point) and/or metering end devices to an NNAP.
- The **M** interface is between this communications function of the meter and the LNAP or between metering end devices. The interface defines the access of external devices to internal data on the meter. The interface profile has to offer services that enable the meter to provide access via the LNAP to the functions implemented in the MID part of the meter or outside it. In BSI terms interface **M** implements a Local Metrological Network (LMN).

Interface	French Linky system	Swiss Landis+Gyr (E450 and DC450)
Interface C	• DLMS/COSEM • UDP or TCP • IPv6 • IETF 6LoWPAN • IEEE 802.15.4 MAC • Mesh ITU-T G3 PLC (power line communication, France's name CPL). PLC-G1 CENELEC A (3–95 kHz). In the future, PLC-G1 will be replaced by PLC-G3 CENELEC A (10–90 kHz)	• DLMS/COSEM + security IEC 62056-5-3 • TCP, UDP • IPv4, IPv6 S-FSK PLC, G3 OFDM PLC, Ethernet, GPRS, UMTS, RF-Mesh
Interface G1	–	–
Interface G2	• DLMS/COSEM • TCP • IP • GSM/GPRS	• DLMS/COSEM • TLS • IPv4, IPv6 RJ45 Ethernet (existing modification with GSM/GPRS/UMTS/LTS)
Interface H1	LAN interface: Euridis (utility interface) 2-way 9600 bps interface. Linky has a physical screen and a remote customer information port. Through this port, users can get various information from the meter (current consumption, apparent power, and tariff period)	An optional integrated wireless interface (ZigBee) enables bi-directional communication with our ecoMeter in-home display. A local port on the smart meter enables consumer access to consumption data. The local port can be implemented by any smart meter physical port e.g. optical port, M-Bus port, IP port (G3 PLC)
Interface H2	Mono-directional line for energy management (7 virtual contacts)—Dry contact (French name "Contact sec") or TIC interface 1-way, 9600 Baud. Exist an option to install on the Linky meter ERL module (Emetteur Radio Linky). This will provide an opportunity to use KNX RF Multi (868 MHz) and ZigBee (2.4 GHz) protocols with the MQTT protocol on the application level. At the meeting, experts from Enedis said that with a special dongle any communication protocol could be used for this purpose	The IDIS CII establishes the foundation for any future consumer HAN services outside the meter. Consumer Interface (open HAN Interface) optical and M-Bus wired + wireless M-Bus (868 MHz). E450 has 1 digital output with a 90 mA relay option and 1 Bistable relay 8A. Also, optionally possible to add 1 Bistable relay 8 A and 1 latching relay 5 A
Interface M		The E450 meter can act as a gateway for collecting data and interacting with other energy meters, like gas, water, or heat. E450 supports DSMR 2.2+ and OMS 4.03. Default E450 has a wireless M-Bus (868 MHz). Wired M-Bus can be used with an additional dongle and use RS485
Interface	German BSI SMGW system	Swiss Siemens (IM150 and SGW1050)

(continued)

(continued)

Interface	French Linky system	Swiss Landis+Gyr (E450 and DC450)
Interface C		• DLMS/COSEM (IEC 62056) • TCP • IPv4, IPv6 • G3-PLC CENELEC A (35–91 kHz) or FCC (150–490 kHz). With external module can be used RS-232 with RJ45 connector, GSM/GPRS/UMTS/LTS
Interface G1	• DLMS/COSEM-IC IEC 62056-6-2 (only COSEM part) • OBIS IEC 62056-6-1 • XML Transfer syntax for COSEM/OBIS objects (optional) • CMS (Cryptographic Message Syntax) IETF RFC 5652 • RESTful COSEM web services (optional) • HTTP 1.1 RFC 7230-7235 • TLS 1.2 RFC 5246 • TCP • IPv4/IPv6 GPRS/EDGE/UMTS/LTE, DSL/Ethernet	• DLMS/COSEM (IEC 62056) • TLS • TCP • IPv4, IPv6 • GSM/GPRS/UMTS/LTS
Interface G2	–	–
Interface H1	• TLS 1.2 RFC 5246 • TCP • IPv4, IPv6 Ethernet > 10 Mbit/s IEEE 802.3i	DSMR P1 interface (RJ12)
Interface H2	• SOCKSv5. Draft RFC "Secure Sockets Layer for SOCKS Version 5" (only for HKS3) • TLS 1.2 RFC 5246 • TCP • IPv4, IPv6 Ethernet > 10 Mbit/s IEEE 802.3i	DSMR P1 interface (RJ12)

(continued)

(continued)

Interface	French Linky system	Swiss Landis+Gyr (E450 and DC450)
Interface M	(a) Wireless bidirectional connection • M-BUS EN 13757-3:2011 • TLS RFC 5246 (M-BUS mode 13) • AFL (Authentication and Fragmentation Level) • Wireless M-Bus EN 13757-4:2011 (b) Wireless unidirectional connection • M-BUS EN 13757-3:2011 • Encryption Mode-7 AES-CBC + CMAC • AFL (Authentication and Fragmentation Level) • Wireless M-Bus EN 13757-4:2011 (c) Wired connection • OBIS IEC 62056-6-1 + DIN EN 13757-1 (OBIS) DLMS/COSEM IEC 62056-6-2 • SML IEC 62056-5-3-8 • TLS 1.2 RFC 5246 • HDLC ISO/IEC 13239 (Format Type 3, CRC according to IEC 62056-46) • EIA/RS-485	Wired and wireless M-Bus (EN13757)

4 Security Analysis

Based on the information provided, it appears that the architecture of French Linky and the Swiss solutions from Landis+Gyr and Siemens are similar from a smart metering systems perspective. Both use the DLMS/COSEM stack of protocols, commonly use PLC G3 on the physical layer, and have a data concentrator in their architectures. The Interoperable Device Interface Specification White Paper (Landis+Gyr White Paper) even points out that Linky smart meters meet IDIS specifications.

In differ to the French and Swiss systems that are considered in this work, the German BSI system uses TLS as the main protocol to secure all 3 SMGW networks. That means to make a comparison from a security protocols point of view for comparison of French, German, and Swiss (Landis+Gyr and Siemens) systems we should compare DLMS/COSEM and TLS (with BSI requirements) protected architectures. We will not describe the TLS protocol in detail because TLS is very widespread and there are lots of materials that describe the main principles of TLS in simplified form.

4.1 DLMS/COSEM Protocol Description

To be more precise, the DLMS/COSEM stack of protocols and vulnerabilities should be considered in more detail. Device Language Message Specification (DLMS) is a server-client protocol that is used for retrieving consumption data from smart meters and for transmission of this data to the energy supplier's Meter Data Management System (MDMS). DLMS works with Companion Specification for Energy Metering (COSEM) and uses OBIS codes (Object Identification System) for meter identification.

COSEM interface classes and their instantiations (objects) are used for modeling energy management use cases including metering. Object modeling is a powerful tool to formally represent simple or complex data. Each aspect of the data is modeled with an attribute. Objects may have several attributes and also methods to perform operations on the attributes. Objects can be used in combinations, to model simple use cases such as register reading, or more complex ones such as tariff and billing schemes or load management. At the moment there are 89 specified interface classes.

OBIS is the naming system of COSEM objects. OBIS codes are specified for electricity, gas, water, heat cost allocators (HCAs), and thermal energy metering, as well as for abstract data that are not related to the energy kind measured. The hierarchical structure of OBIS allows classifying the characteristics of the data e.g., electrical energy, active power, integration, tariff, and billing period.

The DLMS concept was standardized as an international standard by the International Electrotechnical Commission as IEC 62056. IEC 62056 is a set of standards for electricity metering, data exchange for meter reading, tariff, and load control established by the International Electrotechnical Commission (IEC). This series includes the following list of standards:

- IEC 62056-21: Direct local data exchange
- IEC 62056-42: Physical layer services and procedures for connection-oriented asynchronous data exchange
- IEC 62056-46: Data link layer using HDLC protocol
- IEC 62056-47: COSEM transport layers for IPv4 networks
- IEC 62056-53: COSEM Application layer
- IEC 62056-61: Object identification system (OBIS)
- IEC 62056-62: Interface classes.

IEC 62056 standards are focused on electricity metering while DLMS/COSEM is more general and applied to any energy metering. Communication standards differ, e.g., IEC 62056-21 is ASCII-based communication while DLMS is a binary protocol. These standards have been adopted by a large number of manufacturers and service providers making them one of the most widely implemented in smart meters. In the interaction, DLMS/COSEM forms an object model for the monitoring, communication, and transmission of meter readings.

The DLMS/COSEM provides different interface classes to represent real smart metering infrastructure objects and their functionalities. Through the instantiation of these classes, the status and functionality of the measurement equipment are represented, to expose it through the communication network. Physical devices modeled in the protocol contain one or several logical devices that are responsible for modeling the specific functionalities of each device. These objects, as in the object-oriented programming paradigm, are nothing more than a grouping of attributes and methods. Gauges act as servers, which are queried by client applications that retrieve data, provide control information, or perform actions on the gauge through the attributes and methods exposed on the objects defined in the gauge.

To define a meter it is required to define a physical device with its logical devices and instantiate the desired interface classes. The standard defines 70 interface classes that can be used, each with different attributes and available methods. For example, to define an energy measurement, a register class is provided, which will be instantiated by each of the types of measurements that the meter maintains. Taking as an example a meter that is responsible for measuring the consumption of electricity, gas, and water, will be defined as a device containing three logical devices one for each type of energy, and within each one of them will have an instance of the register class, which will contain an attribute with the value of the energy consumption measurement.

The server role is commonly implemented in smart meter devices and the client role on the energy service provider side. To be able to access the objects in a DLMS/COSEM server, it is necessary to establish an Application Association (AA) with the client to identify the participants and establish the context in which the communication will take place, that provides e.g., which authentication mechanism should be used. Servers always have an instance of a special kind of class called an Association, which contains this information and also contains a list of all the objects that are accessible on that particular server so that a client application can know what to do, what information allowed to access, and how to do it.

The protocol provides two ways to access objects, by Logical Name (LN) or by Short Name (SN). Each object always has a LN. In general, manufacturers use the same values in their devices, so that the same application in charge of collecting data from meters can do it regardless of the manufacturer. The LN is the first attribute of a COSEM object and together with the id of the interface class defines the meaning of the object. The LN is defined as a 6-byte OBIS (Object Identification System) code, for example, the association object code is always 0.0.40.0.0.255, and within each object, the attributes and methods are identified with a numeric id.

In the form of access by LN, the methods and attributes are accessed through the id of the interface class, the value of the LN, and the index of the attribute or method which needs to be accessed (class id | logical name | id attribute or method.) In contrast to the SN access form, each object is mapped to just one SN. This is a simplified form of access, indicated to be used by simple devices. In this case, each attribute and method of the objects is identified with a 13-bit integer. Through these two forms, the client can access the objects, read or modify the values of their attributes or trigger actions through their methods. COSEM objects can be accessible by using the address of the client to define which objects that particular client can have access to and of what type (read or write).

For managing the connection between a client and a DLMS/COSEM server, using the services provided by the Association Control Service Element (ACSE). When starting a connection, the client sends an AARQ message to the server, indicating some of its data and what encryption and authentication method it wishes to use, plus additional information such as which version of the protocol to use. In response to this message, the server sends an AARE message, accepting or rejecting the connection attempt, or if HLS is used as the authentication method, partially accepting the connection, waiting for it to complete the extra step of mutual authentication. This extra step consists of the response to a challenge presented by the counterpart. The server sends the client a nonce, whereupon the client sends the server its response based on this nonce, and vice versa. In the case of using MD5 or SHA1, HLS responses are computed based on a nonce and shared secret. When using GMAC to compute the challenge-response, GMAC is applied to the concatenation of a control byte 0x10, the authentication key, and the nonce.

After the connection is established, the client can send requests to the server, both to obtain data and to invoke methods, for which it will receive responses from the server. The messages have a header that indicates the type of message it is and the type of communication. For the termination of the connection, the client sends a server an RLRQ message requesting termination and then the server responds with an RLRE message confirming the conclusion of the communication between them.

In the ISO OSI model, DLMS communicates over L4-L5 (transport and session layer), and COSEM forms the presentation layer (L6) (Matoušek 2017; Table 1):

The DLMS/COSEM protocol allows communication over both TCP or UDP over IP and HDLC networks. In the case of TCP-UDP/IP networks, COSEM services are supported by the COSEM transport layer, which consists of a wrapper over the TCP or UDP protocol that adds an 8-byte header before the message, indicating the version. The protocol used, the source and destination ports, and the length of the message sent. This ensures that the entire message is received before being processed. For DLMS/COSEM, IANA registers port numbers 4059/TCP and 4059/UDP.

Table 1 DLMS/COSEM OSI model layers (European Commission DG Energy 2019)

Layer	Function	DLMS/COSEM
Application	Network process to application	Application
Presentation	Data representation, encryption and decryption, convert machine dependent data to machine independent data	COSEM
Session	Interhost communication, managing sessions between applications	DLMS
Transport	End-to-end connections, reliability and flow control	DLMS
Network	Path determination and logical addressing	DLMS
Data link	Physical addressing	HDLC, IEC 62056-47
Physical	Media, signal and binary transmission	Serial media, cable, radio

4.2 DLMS/COSEM Security

DLMS/COSEM protocol provides two "types" of security—access security and transport security. Access security concerns the rights of a client (for example an application running on a PC) to access data stored on a given server (for example an electricity meter). Transport security concerns the "ciphering" applied to the information exchanged between the server and the client.

DLMS/COSEM protocol defines three authentication security levels:

- Lowest Security Level (NLS)—Neither the client or the server is authenticated.
- Low Level Security (LLS)—Only the client is authenticated by presenting a password to the server. This type of authentication should only be used when the communication is carried out over a secure channel, in order to avoid eavesdropping and message replay.
- High Level Security (HLS)—mutual authentication both client and server authenticate against each other is used. This is the recommended authentication method since in general the security of the communication channel cannot be guaranteed. Possible options—HLS-MD5 (Message Digest 5), HLS-SHA1 (Secure Hash Algorithm), or HLS GMAC (Galois Message Authentication Code).

For transport security DLMS-COSEM defines the concept of 4 different available security policies:

1. Security is not imposed.
2. All messages are authenticated.
3. All messages are encrypted.
4. All messages are authenticated and encrypted.

Moreover, several security suites specify the cryptographic algorithm that is used for message security, such as Security suite 0 which employs AES-GCM-128 for authentication, encryption, and key-wrapping. The major difference between security suite 0 and suites 1 or 2 is that methods of asymmetric cryptography are not available so an offline key agreement is necessary. Security suites 1 and 2 presented in Green Book both use elliptic curve-based digital signatures (ECDSA) and Diffie-Hellman key agreement (ECDH). Additionally, compression can be used (Fig. 8).

DLMS/COSEM defines 3 security suites where AES key wrap (RFC-3394) used for key update (Table 2):

For message encryption, the protocol uses the Galois Counter Mode (GCM) and AES-128 block. Each block in the keystream is calculated using AES-128 based on the 128-bit AES key, a unique initialization vector (IV), and the block counter, which is expressed as a 32-bit unsigned integer that starts with the value 1. It is very important to have a unique IV, since repeating the keystream can have serious consequences, compromising the confidentiality of the communication. The IV in theory can be of any length, but IVs less than 96 bits long are considered unsafe (Dworkin 2007). DLMS/COSEM follows this recommendation and uses as IV the concatenation of the AP Title, a 64-bit value that identifies each device and is exchanged during connection establishment, followed by the 32-bit frame counter.

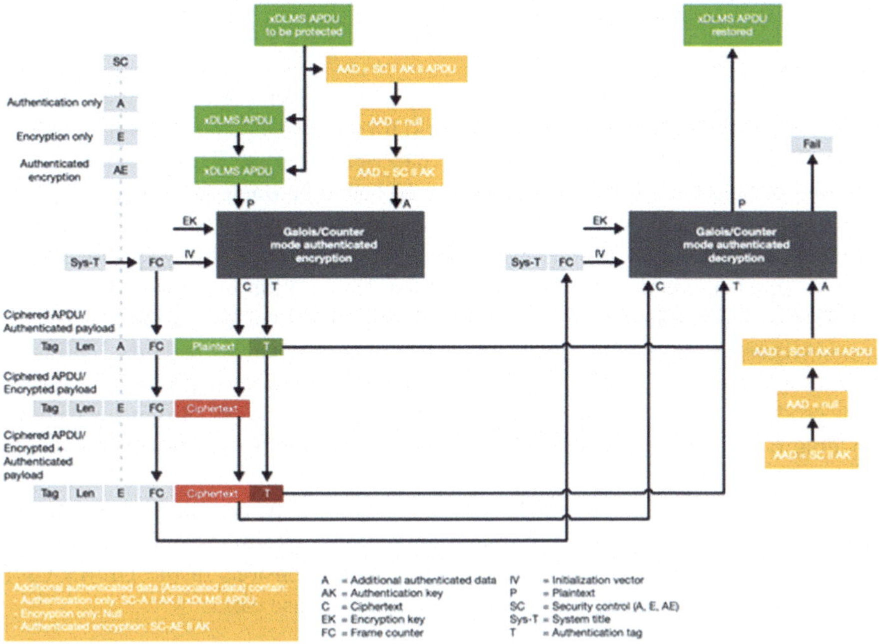

Fig. 8 DLMS-COSEM message cryptography architecture (Landis+Gyr 2014)

Table 2 DLMS/COSEM security suites

Security suite	Authenticated encryption	Digital signature	Key agreement
0	AES-GCM-128	–	–
1	AES-GCM-128	ECDSA with P-256	ECDH with P-256
2	AES-GCM-256	ECDSA with P-384	ECDH with P-384
Security suite	Hash	Key transport	Compression
0	–	AES-Wrap with 128 bit key	–
1	SHA-256	AES-Wrap with 128 bit key	V.44
2	SHA-384	AES-Wrap with 256 bit key	V.44

Transport security is achieved by using ciphered COSEM services (glo_…services) instead of plain COSEM services. All the plain COSEM services (ReadRequest, GetRequest, ReadResponse, WriteResponse, etc.) have a matching ciphered variant (glo_ReadRequest, glo_GetRequest, glo_ReadResponse, glo_WriteResponse, etc.). Ciphered services can only be used within a ciphered application context established by an AssociationRequest "'with ciphering". More concrete info can be found in the source (https://icube.ch/Security/security1.html).

4.3 IDIS Specifications

As Landis+Gyr and Siemens solutions, Linky smart meters (at least some of them) which were described in this document meet IDIS specifications (Landis+Gyr White Paper). IDIS—is Interoperable Device Interface Specification. IDIS in fact, is a tight definition of how the DLMS/COSEM standard has to be implemented. This allows smart meters of different vendors to act interoperable. The IDIS association develops, maintains, and promotes publicly available technical interoperability specifications, based on open standards and supports their implementation in interoperable products. The association manages, administers, and protects the IDIS quality label and supports rigorous interoperability testing to ensure high-quality standards. IDIS is an association for smart metering companies which are committed to providing interoperable products based on open standards. IDIS membership is open to any legal entity providing IDIS conformance-tested equipment. IDIS members include Iskraemeco, Itron, and Landis+Gyr.

The specification team of the IDIS association is currently completing the detailed specifications for IDIS package 1 (based on secured S-FSK PLC communication according to IEC 61334-5-1 considering the latest extensions of the DLMS User Association), supporting the following smart metering use cases:

- Automatic meter registration and system integration
- Remote tariff programming
- On-demand and scheduled meter readings for electricity, gas, heat, and water meters
- Disconnection and reconnection of electricity and gas supply
- System-wide clock synchronization
- Quality of supply supervision at end nodes of the distribution network
- Demand/load management
- Remote firmware update
- Restricted data access to authenticated users
- Secure data exchange by enciphering sensitive information and by authenticating the source of data.

Communication security based on the IDIS DLMS/COSEM standard requires:

- use HLS for authentication security
- use security policy 4 (all messages are authenticated and encrypted)
- Recognized and proven AES encryption, 128-bit key
- 4-stage client-server authentication with GMAC
- Symmetric encryption and keys for meter communication with previously exchanged keys are used to enable good communication performance with embedded devices.

The IDIS security specification references the key scheme of DLMS-COSEM, which is based on a single set of unique symmetric keys per meter. Each key type has a specific purpose (e.g., key encryption, authentication, message encryption) within the DLMS-COSEM communication protocol:

- Master key—is an AES key, programmed into the meter's firmware at the time of manufacturing. It is required to change the encryption method and encryption and authentication keys.
- Encryption key—128 or 256-bit key, used for encryption of DLMS/COSEM messages.

Authentication key—It is a key of the same length as the encryption key. It is used to calculate the authentication tag of the messages sent, this authentication tag serves as proof of the integrity of the message and demonstrates the knowledge of the authentication key by the sender of the message (Table 3).

Table 3 DLMS/COSEM key types (Landis+Gyr 2014)

Key type	Use	Generation
Master key (MK)	Key encryption key for global keys	Landis+Gyr production system
Global unicast encryption key (GUK)	Global encryption of unicast xDLMS APDUs	HES
Global broadcast encryption key (GBEK)	Global encryption of broadcast xDLMS APDUs	HES
Global broadcast authentication key (GAK)	Authentication of xDLMS APDUs	HES
Dedicated unicast encryption key (DUEK)		
Key typc	Delivery	Location
Master key (MK)	From Landis+Gyr production system to HES using signed messaging	• Production system • HES • Meter
Global unicast encryption key (GUK)	Wrapped with master key, invocation of global_key_transfer method by HES or the DC	• HES • DC • Meter
Global broadcast encryption key (GBEK)	Wrapped with master key, invocation of global_key_transfer method by HES or the DC	• HES • DC • Meter
Global broadcast authentication key (GAK)	Wrapped with master key, invocation of global_key_transfer method by HES or the DC	• HES • DC • Meter
Dedicated unicast encryption key (DUEK)	Transported as part of the xDLMS Initiate Request APDU, which is encrypted and authenticated using the AES-GCM-128 algorithm, the global unicast encryption key and the authentication key	

To gain a deeper understanding of the security key generation and management possible to consider the Gridstream IDIS process (Landis+Gyr 2014):

1. The Landis+Gyr manufacturing facility uses secure key management software and secure key storage hardware to generate an initial unique key set for each meter consisting of a master key (encryption key) and initial global keys (GUK, GBEK, and GUEK). It is important to note that the DLMS term "global" means valid over multiple communication associations, and not system-wide.
2. The global keys are then encrypted using the master key and written to the meter.
3. The Landis+Gyr production system sends a copy of the key material to the utility AIM system using signed secure based on the Landis+Gyr public key infrastructure.
4. The utility AIM system stores and manages the key material using its local secure key manager and secure key storage hardware.
5. As each meter is registered to the AIM system as part of the installation process, AIM securely distributes the key material to the appropriate data concentrator (using TLS over mobile communications) and initiates communication with the meter.
6. As a part of the communication initialization process, AIM renews the meter's global keys and distributes them to the data concentrator and meter.

All communication from the head end system to the meter via the data collection system is authenticated and encrypted using the renewed meter-specific keys.

4.4 DLMS/COSEM Vulnerabilities

The DLMS/COSEM protocol is not free of vulnerabilities, both the protocol itself and particular to its different implementations. Below listed some of the known DLMS protocol vulnerabilities (Lüring et al. 2018; https://research.tue.nl/en/studentTheses/dlms-cosem-protocol-security-evaluation):

- Optional authentication—The use of authentication is optional according to the definition of the protocol, and it is independent of the encryption of the messages. An attacker can manipulate the security byte of messages and truncate the message in such a way as to remove the authentication byte and thus maintain an unauthenticated communication, without the receiver knowing if this was the case, the original intention of the issuer.
- Information leaking—each message contains a header indicating the type of message and communication mode used. This is unnecessary since the type of message is revealed even when it is an encrypted communication, making it possible for attackers to know what information is being transmitted. This could be replaced by simply indicating whether the communication is encrypted using the global key or a dedicated key and the indicated message type encrypted with the rest of the message. Devices do not require this additional information in order

to decrypt the message. But even in this case, because each service has a fixed, well-known, preamble and message structure an attacker may be able to perform a known plaintext attack.
- Vulnerable Authentication Methods—In HLS, the method used to authenticate the client and the server are the same. Given the right circumstances, an attacker could impersonate a valid server by replaying the AP title, nonce, and the response to the authentication challenge. A rogue server may reply to the client CtoS and f(CtoS) to trick the client that he knows the secret key. Since the f(CtoS) and f(StoC) are exchanged using the execute service, the attack requires that the APDUs are exchanged in plain text. To prevent it—the client must reject association responses if StoC is equal to CtoS.
- When HLS association is performed using MD5 or SHA1, offline dictionary attacks are possible, since the nonce (sent in plain text), the challenge-response, and the authentication function are known. With this information, given enough time and resources, the key used can be calculated. If an adversary acquires a valid HLS response and its corresponding nonce (e.g., via server impersonation or by sniffing traffic) he can then try to find out the shared secret. An attacker has nonce and $h = f$ (nonce || password). Offline he can try several passwords until he obtains h. To prevent this, the use of MD5 or SHA1 mechanisms should be forbidden, and required to use of randomly generated secrets.
- Possibility of injecting answers to the client—The responses are not linked to the requests, that is, given a request sent by the client, the responses can be replaced by another message and as long as it contains the expected data type, it will be taken as a valid response.
- Security downgrade—even with high security mode using of authentication mechanism is non obligatorily. XOR keystream model ciphers are vulnerable to bit-flip attacks, which makes it possible to remove message authentication tags. This vulnerability of DLMS/COSEM protocol leads to the possibility of acquiring ciphertexts by known parts of plaintext, authentication mechanism disables, and message replacement.

DLMS/COSEM implementations are also not free of bugs. Some vulnerabilities that may be present in particular implementations of the DLMS/COSEM protocol are listed below (https://research.tue.nl/en/studentTheses/dlms-cosem-protocol-security-evaluation):

- Invocation Counters Unenforced—the server accepts messages whose frame counter is less than or equal to the counter of the last encrypted message received from a device. This results in a lack of adequate protection against replay attacks.
- Predictable Association Challenges/Nonces—If predictable nonces are used, an attacker could calculate the necessary HLS responses to be able to authenticate with the server or client. Highly unpredictable nonces shall be used (e.g. using a CSPRNG or a TRNG). As a solution—implementations should not use a linear congruential generator.

- Identical AP Titles Allowed—If communication between two devices with the same AP title is allowed, a client or server can allow communication with a DLMS/COSEM device with the same AP title. This could allow replay attacks.
- Arbitrary System Titles Accepted—For each different ST the last used IC shall be remembered. If arbitrary ST are accepted, an attacker may attempt a Denial of Service (DoS) attack by filling the IC database. If a ring buffer is used, an attacker may attempt to reset one counter by filling up the buffer and eventually proceed with a replay attack.
- The messages are processed only based on the header that indicates the type of message, without verifying their content—If messages are processed only taking into account the header indicating the type of message, an attacker could send a message whose type indicates that it is an encrypted message, but with the encryption and authentication bits disabled in the byte of security, potentially tricking the server into processing a malicious message sent in plain text.
- Encrypted AARE messages are sent in response to plaintext AARQ—This may allow this information to be used to obtain the keystream since the values of the AARE messages are known based on the AARQ sent.
- Possibility of online dictionary attacks—Multiple authentication attempts are allowed, allowing online dictionary attacks. After N connection attempts, the error messages should be the same.
- Messages with an invalid authentication tag are allowed—If messages are processed without checking the authentication tag, the messages are vulnerable to integrity violations.
- Encrypted messages without authentication are allowed—Encrypted messages without authentication are susceptible to manipulation, so it is desirable that they are not processed.
- Insecure authentication methods are allowed—LLS should not be used as the password is sent in plain text. In addition, proprietary authentication methods, and the MD5 and SHA1 versions of HLS are vulnerable to man-in-the-middle attacks.
- Ciphered APDU Type Ignored—when in the security header the tag leaks information about the secured message type (e.g. Get), the contained plain text message type shall be consistent. Some devices ignore this and accept the message.
- Plain Text APDU Accepted—some implementations that are supposed to accept only secured messages can be fooled to accept plain text messages by simply using the security header with both the crypto/auth bits turned off.
- Message authentication code (MAC) not enforced—Messages with invalid MAC are accepted. Invocation Counter Reset After Reboot On power loss, some implementations reset the IC to zero.
- Premature Session Termination—HLS Associations shall terminate with an encrypted termination message. Some implementations accept plain text termination messages, thus allowing an attacker to disconnect legitimate sessions.
- Default keys on production—Meters on the field are occasionally left with their manufacturer default keys. The keys are not only equal between user profiles, but also between several hundreds of meters.

- Client Skips HLS Authentication Check—some clients do not check rogue servers and just ignore the received f(CtoS) response.

Covering all of the considered vulnerabilities is possible via adding additional restrictions to DLMS/COSEM software implementation. As an example—possible to make the implementation force reject a message if this message does not use an authentication mechanism. By information that we have from SMI project meetings and direct responses—Landis+Gyr, Siemens SMI, and Linky solutions should be resilient to described vulnerabilities because they are implemented according to the IDIS specifications.

4.5 TLS with BSI Restrictions and DLMS/COSEM Comparison

Transport Layer Security (TLS, formerly Secure Sockets Layer or SSL), is a cryptographic protocol designed to provide secure communications over computer networks. TLS could be represented as the special shell that provides encryption and integrity (to prevent eavesdropping and tampering) of higher-level data protocols such as email, instant messaging, voice over IP, or HTTP. TLS was proposed by Internet Engineering Task Force (IETF) as a standard and was first defined in 1999 on the earlier SSL protocol specifications (1994, 1995, 1996) developed by Netscape for adding secure HTTP protocol to their Navigator web browser. The modern TLS version 1.3 was defined in August 2018, but BSI SMGW documentation requires to use of TLS version 1.2 whose implementations are currently much better tested compared to version 1.3. The TLS protocol includes two main parts—the TLS record and the TLS handshake protocol. When a TLS server and a TLS client have agreed to use TLS, they use a handshake with asymmetric cryptography to establish cipher settings and create a session-specific shared key which is later used for symmetrically encrypted communication. The handshake protocol is responsible for choosing parameters that will be used to establish the secured connection.

Comparing TLS and DLMS/COSEM stack of protocols, they have similar cryptographic security levels. Both use PKI (Public Key Infrastructure), modern security primitives, and cipher suites. The difference between DLMS/COSEM and BSI SMGW TLS PKI is that in the SMGW PKI uses a state-controlled root-CA which means that the signature should be legally standardized and issued certificates have demanded lifetime periods for all three network interfaces in TR (Table 4).

The main difference between these two protocols is that DLMS/COSEM is less complex compared to TLS. TLS can be regarded as a "wrapper" protocol that is used to secure application-level data, whereas DLMS/COSEM is a more specialized protocol which theoretically should reduce the chance of vulnerability appearance. TLS is a more complicated protocol which means that TLS has more points that potentially could be vulnerable to an attack. The TLS protocol is designed to be

Table 4 TLS and DLMS comparison (Lüring et al. 2018)

	DLMS/COSEM	BSI SMGW TLS
PKI	+	+
Authenticated encryption	AES-GSM-128 AES-GSM-256	AES-GSM-128 AES-GSM-256 AES-CBC-128 AES-CBC-256
Elliptic curves	NIST P-256 NIST P-384	NIST P-256 NIST P-384 BrainpoolP256r1 BrainpoolP384r1 BrainpoolP512r1
Digital signature	ECDSA	ECDSA
Key agreement	ECDH	ECDHE
Key transport	AES key wrap	AES key wrap
Hash function	SHA-256 SHA-384	SHA-256 SHA-384
Message authentication code	GMAC	CMAC

flexible and adaptable, allowing for the easy addition or removal of security primitives and algorithms as needed, such as in the case of a discovered vulnerability. The TLS protocol is also designed to be independent of any specific security primitive or algorithm, and there are currently over 442 cipher suites registered with the IANA that can be used with TLS implementations. In the case of DLMS/COSEM implementations changing the algorithm will be more complicated in comparison with TLS, because currently DLMS/COSEM supports only 3 security suites. Also, because DLMS/COSEM supports only 3 Security Suites—implementations are more lightweight, which is essential for embedded systems.

One more advantage of the TLS protocol is its widely carried out research for potential vulnerabilities. The reason for this lies in the widespread use (e.g., in banking and shopping businesses areas where secure Internet connections are very important). There are significantly more known vulnerabilities in the TLS protocol compared to DLMS/COSEM, with 781 reported for TLS compared to just 2 CVEs for DLMS/COSEM. CVE (Common Vulnerabilities and Exposures) is a system for publicly known information-security vulnerabilities. However, there appears to be significantly more research and testing conducted on TLS, as indicated by the higher number of papers on the topic in Google Scholar, with over 25,000 papers published on TLS compared to just 862 for DLMS. DLMS has info only about 2 CVEs and both in free open source software Gurux.

Two more differences are the used random number generation mechanisms and secure key storage. In the case of BSI SMGW, TLS TR requires a certain number of generator classes that can be used and cryptographic materials should be stored in a special SMGW security module. In the case of DLMS/COSEM the specification just requires the use of a "strong random number generator" and does not have any requirements for special cryptographic materials storage.

5 Theoretical SMI Systems Security Comparison

To compare French, German, and Swiss smart metering infrastructure we start with a comparison of the supervisory authority's requirements for smart metering infrastructure solutions. France had very concrete legal data protection since 1978 (French National Assembly and the Senate 1978), but technical requirements for smart metering systems are the most abstract compared to German and Swiss supervisory authority's requirements. Due to Enedis energy distribution market dominance, the problem with lack of technical requirements is partially solved. Because Enedis managed to create a great system with a lot of useful smart metering features and for comparison purposes, we use a Linky system as the only French smart metering solution for comparison. One of the cons of the Enedis documentation is that most Linky features are only described in marketing materials (such as Enedis presentations/brochures; https://www.enedis.fr/documents?term_node_tid_depth%5B106%5D=106) and we did not find much publicly available technical documentation (such complete as German TR-03109-1). Some information about the Linky system was received from Enedis speakers at project meetings, but unfortunately, we were not able to find some documentation proof and should make conclusions based only on the speaker's words and marketing materials. In the document "Linky PLC profile functional specifications" [ERDF-CPT-Linky-SPEC-FONC-CPL (EDRF 2009)] Enedis refers to DLMS/COSEM (IEC 62056) protocol from "Coloured books", in which authentication and encryption are used as the main Linky system security mechanisms. From a security point of view, the Enedis energy market-dominating has some advantages. Using only one device type for a smart metering solution means that the algorithm for making any device settings is also the same. The probability that service technicians will do incorrect settings that could lead to a system vulnerability is lower than in the case of multiple solutions from different models.

BSI creates a complete documentation system that fully describes an architecture, protocols, access rights, and other requirements for a smart metering infrastructure. All documentation is publicly available on the BSI website (https://www.bsi.bund.de/DE/Themen/Unternehmen-und-Organisationen/Standards-und-Zertifizierung/Smart-metering/Smart-Meter-Gateway/smart-meter-gateway_node.html). In contrast to France, in the German distribution network exist about 883 distribution companies of various sizes. It is likely that the decision to have a longer process for developing the requirements and documentation structure for the German BSI smart metering system contributed to the fact that it was implemented later than the systems in France and Switzerland. BSI documentation gives the impression of transparency and a comprehensive elaboration in comparison with Swiss and French SMI supervisory authorities' requirements.

But different from the German system, Swiss supervisory restrictions in the field of architecture and used protocols are very low. Theoretically, that can lead to an increase in the probability of vulnerability presence because of the large number of used protocols and device variations. From another point of view, if a significant vulnerability is found in one system, for example in Landis+Gyr, that will not mean that the Siemens system will have the same problems, and the probability that all smart metering solutions in the country will be vulnerable in one moment is lower than in French system. Landis+Gyr and Siemens provide answers to most of our questions about their systems but were careful in questions about the security part.

Based on information from the previous sections it is possible to say that from the smart metering systems architecture point of view French Linky and Swiss solutions from Landis+Gyr and Siemens are very similar. These systems include several points that theoretically could be attacked:

1. Energy service provider backend system
2. WAN connection from an energy service provider to a data concentrator
3. Data concentrator device
4. PLC connection from a data concentrator to a smart meter
5. Smart meter device
6. Connection from smart meter device to controllable local devices
7. Controllable local devices.

German BSI SMGW system smart metering architecture is slightly different from the solutions described above. Describing the points on which it is theoretically possible to make an attack, possible to distinguish:

1. Energy service provider backend system
2. Smart meter gateway administrator system
3. Smart meter gateway operator system
4. WAN connection from SMGW to external market participants
5. SMGW device
6. LMN connection from SMGW to meter devices
7. Connection from smart meter device to controllable local devices
8. Controllable local devices.

To make the analysis more correct it is important to divide the attackers into two groups—internal attackers and external attackers. Internal attackers are household owners who are interested in energy consumption data tampering to reduce their costs. They are not interested in controllable local device hacks because they already have physical access to them. The role of outside attackers can perform as malefactors which want to gain profit or potential adversary services interested in causing damage to destabilize the local situation.

Because of very soft restrictions for Swiss smart metering systems, both German and French SMI systems potentially can be used in Switzerland. All systems considered in this report use the OBIS model, which means that theoretically, they could use the same software in the backend system.

From the point of view of an outside attacker, the most interesting aim is the energy service providers backend system. We do not have the permission to make energy service provides backend systems scanning or penetration testing in this project. Therefore, we do not have information about services and software running on the energy providers servers, but we are sure that at least it should have a service to receive consumption data and control home appliances on the energy consumer side (in the case of France and Germany). Enedis has a web API that can be used by users to collect consumption data (https://data.enedis.fr/api/v2/console). The probability that the same company network will run web, FTP, SSH, or some other common services that could be vulnerable and used as entering points is higher than the probability of finding a vulnerability in other elements of smart metering architecture because all devices considered in this work are relatively simple in terms of operating logic compared to the backend part and have a small number of "influence points" which are possible to test.

Also, the hacking of backend servers provides a large list of opportunities that cause the most damage to the system such as:

- Accessing the consumption data of a large number of users.
- Gaining access to smart metering devices control (such as data concentrators or smart meters) which, for example, can be used to create a botnet or simple devices deny of service.
- Make a "blackout" by getting remote access to power substations.

Some artificial intelligence and predictive maintenance tools such as Enedis Carto-Line theoretically can be used to make some harm to the systems. Knowledge about calculations algorithms and access to manipulating smart metering device consumption data or data concentrator should be possible to use for transmitting incorrect data to HES, which will lead to the incorrect settings on electricity substation and can therefore cause a blackout or overvoltage which could in turn lead to a system running out of order. Also, stopping all the meters at the same time could create an energy excess and a break in the network.

To make a botnet it is important to have the ability to upgrade firmware or install additional software. At the SMI project meeting, the speaker from Enedis said that a Linky meter can produce a remote and local firmware upgrade. In open sources, there was no concrete information about how the Linky, Landis+Gyr, or Siemens devices firmware upgrade process is organized (how to transfer a new firmware image to the device and how to secure this process). But there exists information about how the firmware upgrade process could be done with smart meters that use DLMS protocol in the documentation of open-source software for DLMS-based smart meter managing—Gurux (https://www.gurux.fi/front-page). Based on the information from the Gurux documentation, it is required to use the "highest" security mode of the DLMS connection (with ciphering, authentication, and encryption) to perform a firmware upgrade.

In contrast to the French and Swiss Landis+Gyr and Siemens solutions, in the case a hacker gets access to the German BSI energy provider backend system—a hacker will only get access to the store on this server energy consumption data and to the controllable local devices, which were previously confirmed by the smart meter gateway administrator. Only SMGW administrators can install firmware updates or add a new controllable local system devices. The SMGW administrators should check for firmware updates from SMGW producers' resources and only after successful verification install updates on SMGWs.

One other important aspect is the SMGW's "wake up" service—which is described in the "German smart metering infrastructure" section. Because of the "wake up" service hackers will not even be able to scan a SMGW from the WAN side even if a SMGW will have a public IP address in the WAN and therefore the task to hack a SMGW from the WAN side looks close to be impossible. Even if an attacker will be able to bypass the wake-up service, the SMGW will initiate a TLS connection to the SMGW administrator, which will be protected by TLS and also requires to be hacked to make some harm. In case of DLMS/COSEM based devices, there is no information about such things as BSI "wake-up" service, and because of that in the systems where smart metering devices implement the DLMS/COSEM server role, an attacker can perform any scan or attack attempt if he is able to get access to the required network.

Linky does not have an additional protection by TLS on a connection from the data concentrator to the Energy service provider backend system and only uses the PLC protocol at the physical layer (EDRF 2009). Was found the source which mentions the use of TLS in the context of the Linky system—"The Linky system uses more traditional protocols, not specific to the IoT world, such as https, the ins, and outs of which are better understood. It was thus possible to rely on TLS to secure exchanges" (Marcellin 2018). But in this case, the speaker probably means some Enedis website, because none of the documentation or some promotion materials contains info about TLS.

The PLC G3, which is used in the French Linky, Swiss Landis+Gyr, and Siemens systems to connect a smart meter to a data concentrator, uses AES128-CCM encryption for Data Link (OSI layer 2) security (Genest et al.). For the German BSI SMGW system, the most common WAN type is a cellular network LTE connection. In the case of LTE data encryption exists 4 possible options (https://www.sharetechnote.com/html/Handbook_LTE_EEA.html.; Bartock et al.):

1. 0000—Null ciphering algorithm
2. 0001—SNOW 3G—stream cipher designed by Lund University (Sweden)
3. 0002—AES128—Block cipher standardized by NIST (USA)
4. 0003—ZUC—stream cipher designed by the Chinese Academy of Sciences (China).

As we can see, both options are equally protected at the data link level, but the LTE solution has more possible options.

One more possible direction of comparison is the use of HSMs (Hardware Security Modules). In the case of the Linky system, there is not much information about HSM. It is just one time mentioned in the source (Marcellin 2018), where said that HSM "allows data to be stored" and on page 9 of the source (Nguyen) is shown that HSM is used in the data concentrator device.

German BSI Hardware security module is a different physical device, which allows to divide security and measuring functionalities. BSI TR-03109-1 strictly defines how HSMs should be used. A BSI SMGW must use a HSM for the TLS handshake and other cryptographic operations. For mutual authentication between a SMGW and meters in the LMN, LMN certificates which are X.509 self-signed certificates must be used.

Landis+Gyr Gridstream solutions include the next information about the HSM module (Landis+Gyr 2020)—the HSM serves as the root of trust where the utility ECC private key is vaulted. The private key is used to generate digital signatures to downstream commands sent by the HES. The HSM also features FIPS 140-2 and Common Criteria Level 4 certifications, providing strong protection for one of the critical elements in the advanced security architecture.

In the case of Siemens solutions were found such information as (Jöbstl 2019)—Siemens is using dedicated appliances such as the Hardware Security Module. This module is responsible for the whole meter data decryption and is the only system on which clear text meter keys are present. Meter keys never leave the tamper-resistant foil-coated CPU or its cache. This ensures a physical protection layer to the smart meter data security.

Regarding other architectural aspects, commonly data concentrators (which are used in Landis+Gyr, Siemens, and French SMI systems) are installed in the public territory and have an Ethernet interface with a web server to make settings. Because this element is installed outside the household, compared to the German system, this adds an additional potential vulnerable point to the system, which theoretically can be attacked if the malefactor has physical access from the smart meter to the data concentrator side (for example adding noise to PLC connection in the case to organize DOS attack) and from WAN side.

The German BSI SMGW is commonly installed inside a household. In case a wired connection is used for the LAN and the HAN network is not connected to the home internet router, then the only way to get access to this system from the outside is the WAN (which is well protected as described above). Theoretically, a malefactor can try to attack a SMGW administrator, but due to the limited number of tasks and complicated certification conditions success probability of this attack is quite low compared to an attack on an energy service provider network or data concentrator in French and Swiss Landis+Gyr and Siemens systems. An implementation of passive data sniffing (e.g., if a malefactor will get access to the mobile network base station), will be more complicated in case of the German BSI SMGW, because the consumption data is additionally protected by TLS on the CMS level. Even SMGW administrators could not read energy consumption data, because only the energy distribution service has a key to decrypt the data on the CMS layer.

From an architectural point of view, French and Swiss architectures are theoretically more vulnerable to outside attackers compared to German systems, but less vulnerable to the inside attacker, because the electricity meter module is placed in the same case as the communication unit. In Germany, it is allowed for the meter and SMGW modules to be placed in the same box "SMGW-PP 1.4.5.3 Possible TOE Design: One Box Solution", but there are no known devices on the market that use this architecture. Both German and French SMI systems may potentially be used in Switzerland due to the relatively lenient regulations for Swiss smart metering systems.

Comparison of smart metering systems parts installed inside the household is complicated. In the case of French and Swiss systems, by information that we got from Enedis, Landis+Gyr and Siemens any communication protocol could be used by connecting an additional dongle to the main smart meter system. There are no restrictions in documentation from supervisory authorities on this part. Based on this information, the only conclusion that we could do is that potentially these systems could not be protected at all (for example in the case with wrong settings) and are very vulnerable because of this. From a security point of view—one more advantage of Landis+Gyr and Siemens and French Linky system is that the energy meter and communication module are placed at the same device—which makes internal attacker's actions more complicated because there are fewer entry points in comparison to the SMGW system.

In case of the German BSI SMGW, an attack on the local metrological network (that connects meters and SMGW) will be problematic to an outside attacker, because the SMGW is commonly installed inside the household and in case of a wired LMN connection the SMGW device and energy meter device are installed in the same box. BSI allows several protocol variations for LMN and WAN, it increases the chance of misconfigurations. As we know from our partner the local DSO configuration settings process of Ethernet, mobile network for WAN connections, wireless M-Bus or wired connection for LAN differ a lot. Wireless M-Bus short connection range makes the possibility of a massive SMGW devices DoS (denial of service) attack on LMN extremely difficult for an outside attacker and the probability that an outside attacker will be interested in LMN data tampering or DoS attack only on one device is very low.

6 Use Case: Secure Smart Meter Gateway

As we can see from previous sections, the German BSI SMGW has a very well elaborated security system. Currently, BSI has certified 4 SMGWs (SMARTY IQ, CASA 1.0, PPC LTE, CONEXA 3.0), and 5 more SMGWs are currently under the certification process. Even SMGW device acquisition is a complicated task.

One of the possible alternatives is to use some virtual SMGW. But the only project that was found is jOSEF (A Java-Based Open-Source Smart Meter Gateway Experimentation Framework). Unfortunately, the jOSEF project does not implement SMGW security features responsible for encryption and authentication (TLS, PKI, etc.).

To acquire a real SMGW device, it is required to register as a system operator (Anlagenbetreiber) at German Federal Network Agency (BNetzA) website www.marktstammdatenregister.de/MaStR/. IvESK got a license MaStR-Nummer: ABR931965228164. After that, we get the permission from the local DSO company to test two SMGW devices CONEXA 3.0 from Theben and smart meter gateway PPC LTE from Power Plus Communications AG. But there are a number of limitations to testing these devices:

1. We should not try to test SMGW admin and EMP infrastructure.
2. We should not do any physical harm to devices (including device disassembly to try to download the firmware or affect some interfaces).

In general, it means that we were only able to test SMGW device interfaces in form of a "black box" testing, because we do not have some shell running on the device or even device firmware. During the information gathering and reconnaissance phase the SMGW firmware of PPC was found on their website (https://gwafirmware.ppc-ag.de/). But the download from the website requires authorization.

Device Scanning and Information Gathering

First, we needed to find a way to communicate or interfere with the testing devices and to find SMGW ports available for interaction. For this purpose, the Netdiscover software was used for active/passive address reconnaissance by actively sending ARP requests.

The Theben CONEXA 3.0 SMGW box has three Ethernet ports (Table 5). The WAN-1 Ethernet port is probably turned off because Netdiscover shows nothing on this port. Even when the Ethernet port is connected to the adapter port the LEDs do not start to blink. It is most probable that this port has been completely disabled by the administrator because the SMGW uses a cellular network for a WAN connection.

The CONEXA 3.0 SMGW and PPC SMGW show different MAC addresses in CLS and HAN Ethernet ports. However, the IP address is static and in both ports and set to192.168.100.100. The CLS Ethernet port does not respond to ping requests. Only the HAN Ethernet port on PPC SMGW can be reached.

Finding the IP addresses in the HAN for both SMGWs was successful. Both SMGWs have static IP addresses, but all other parameters are different. In the case of the CONEXA 3.0 SMGW, it has an IP address 192.168.0.1 on both HAN Ethernet ports that have the same MAC address (one on the device corpse and one in the

Table 5 Available SMGW's interfaces

CONEXA 3.0 SMGW	SMGW PPC LTE
• 1 WAN Ethernet port (RJ45) • 1 WAN LTE antenna port (FAKRA SMB connector) • 1 SIM card slot • 1 CLS (HAN) over Ethernet port (RJ45) • 1 HAN over Ethernet port (RJ45 on MTG Mehrwert Konnektor module) • 1 LMN over RS485 (RJ12) port • 1 LMN Wireless M-Bus (OMS) antenna port (FAKRA SMB connector) • Power supply connector	• 1 WAN LTE antenna port (FAKRA SMB connector) • SIM card slot • 1 CLS (HAN) over Ethernet port (RJ45) • 1 (HAN) over Ethernet port (RJ45) • 1 LMN over RS485 (RJ12) port • 1 LMN Wireless M-Bus (OMS) antenna port (FAKRA SMB connector) • Power supply connecter

Mehrwert module), which means that probably the Mehrwert module works as a simple switch. In case of the SMGW PPC the LTE HAN Ethernet port and CLS (HAN) port have the same IP address 192.168.100.100, but different MAC addresses which differ only in the last bit.

For finding running services the zenmap port scanner software was used. Zenmap is the official Nmap Security Scanner GUI.

List 1 SMGW Conexa 3.0 HAN zenmap scan

```
PORT    STATE SERVICE    VERSION
443/tcp open  ssl/https?
| ssl-cert: Subject: common-
Name=ETHE03XXX.SMGW/organizationName=SMGW/countryName=DE
| Subject Alternative Name: DNS:ethe03XXX.sm
| Issuer: common-
Name=ETHE03XXX.SMGW/organizationName=SMGW/countryName=DE
| Public Key type: unknown
| Public Key bits: 384
| Signature Algorithm: ecdsa-with-SHA256
| Not valid before: 2021-01-22T09:08:00
| Not valid after:  2026-01-22T09:08:00
| MD5:   e31e e6eb 1089 4991 46b0 f839 6d85 5781
|_SHA-1: f94f 7a10 6716 8102 fcbf 4cb2 05b6 501e 6d6e 63d9

1080/tcp open  socks?

MAC Address: 5C:CA:32:XX:XX:XX (Theben AG)

Warning: OSScan results may be unreliable because we could not find at least 1 open
and 1 closed port

Device type: general purpose
Running: Linux 3.X|4.X
OS CPE: cpe:/o:linux:linux_kernel:3 cpe:/o:linux:linux_kernel:4
OS details: Linux 3.10 - 4.11, Linux 3.2 - 4.9
```

As presented in the list above, the default zenmap scanner only shows two open ports in the Conexa SMGW HAN. It is a TLS server that runs on default port 443 which fulfills the requirements of BSI TR-03109-1 interface realization IF_GW_CON CON for HAF1—end-user data provision and, probably, HAF2: Service technician data provision. This TLS server uses a fairly reliable signature algorithm mechanism ecdsa-with-SHA256. The fingerprinting NSE script cannot define the CONEXA SMGW HAN TLS server version.

The second open port is 1080, which zenmap defines as socks. This port is responsible for the implementation of HKS3 (home communication scenario 3—transparent channel initiated by CLS). One more interesting observation is that sometimes this socks port just disappears even when the scan was done exactly with the same parameters. Unfortunately, we did not find a reason for this behavior.

Another important moment is defined by zenmap in the OS details section Linux version, but it will be described later.

List 2 SMGW PPC LTE HAN zenmap scan

```
PORT     STATE  SERVICE        VERSION
443/tcp  open   ssl/http       lighttpd
| ssl-cert: Subject: commonName=EPPC02XXX/organizationName=OpenLimit-
PPC/countryName=DE
| Subject Alternative Name: othername:<unsupported>
| Issuer: commonName=EPPC02XXX/organizationName=OpenLimit-
PPC/countryName=DE
| Public Key type: unknown
| Public Key bits: 256
| Signature Algorithm: ecdsa-with-SHA256
| Not valid before: 2020-03-04T10:47:53
| Not valid after:  2027-03-04T10:47:53
| MD5:   0421 1112 0256 bcf5 6ab9 c7cc 185a 9ae7
|_SHA-1: 1f53 f675 84e9 9351 e6b2 6d38 8d50 4d78 1bf6 63d1
2222/tcp open   ssh            Dropbear sshd 2017.75 (protocol 2.0)
8001/tcp closed vcom-tunnel
8002/tcp closed teradataordbms
8003/tcp closed mcreport

MAC Address: 00:25:18:XX:XX:XX (Power Plus Communications AG)

Aggressive OS guesses: Linux 3.10 - 4.11 (98%), Linux 3.2 - 4.9 (94%), Asus RT-
AC66U WAP (93%), Linux 4.4 (92%), Linux 4.1 (92%), HP P2000 G3 NAS device
(91%), Linux 3.16 - 4.6 (91%), Linux 2.6.32 - 3.13 (91%), OpenWrt Chaos Calmer
15.05 (Linux 3.18) or Designated Driver (Linux 4.1 or 4.4) (90%), Android 5.1 (90%)

No exact OS matches for host (test conditions non-ideal).
```

Default zenmap monitor shows 2 open ports in PPC SMGW LTE HAN. Regarding TLS port 443 it implements the same like with CONEXA 3.0 SMGW, but in this case zenmap was able to fingerprint web server software—lighttpd. Other open port is SSH server which listen port 2222. Zenmap was also able to fingerprint version of SSH server—is Dropbear sshd 2017.75.

Also, zenmap define ports 8001, 8002 and 8003 as closed. In zenmap is some port is marked as closed this means that the port is accessible (it receives and responds to Nmap probe packets), but there is no application listening on it (Nguyen). During our research we did not find any information regarding purpose of this ports.

Possible Attack Directions

After the analysis of our theoretical comparison section and the scanning of real devices, we define a list of possible SMGWs testing directions:

1. The direct attack to the energy service provider network
2. The direct attack on the SMGW administrator network
3. SMGW manufacturer website firmware modification
4. Use registered by SMGW administrator CLS certificate to penetrate the energy service provider network via TLS tunnel
5. WAN PKI attacks
6. WAN Wake-Up service fuzzing
7. Manual Wake-Up package overflows of header identification, recipient, and timestamp fields
8. WAN port DDOS. Because in PP 1.4.6.5 step 1.d check wake-up message has not been received before, and only after that check the signature of the wake-up message. If the number of these wrong messages will be big enough, before SMGW will check the real SMGW admin wake-up package, the timestamp of this package already will be outdated
9. "Received before" packages logs overflow
10. LMN AFL tests
11. LMN consumption data tampering (TLS attacks)
12. Internal logs overflow (consumption, actions, etc.)
13. HAN TLS server common vulnerabilities testing
14. HAN TLS server fuzzing
15. HAN username/password sniffing
16. HAN Web servers testing
17. CONEXA SMGW HAN socks server testing
18. PPC SMGW HAN SSH server testing.

Testing of 1–4 was unavailable for us because of the testing agreement conditions and lack connected to SMGW CLS devices to be able to test it. 5–9 requires to have an access to the SMGW WAN port. Because the wired WAN option was disabled, we attempt to interact with SMGW using a Wideband Radio Communication Tester R&S CMW500 and create a MITM connection. For this purpose, we use a spectrum analyzer to find the LTE Band which the SMGW uses to communicate via the cellular network, change the SIM card to a special one and wirelessly connect the SMGW to the CMW500. We succeeded in LTE-level traffic sniffing but were unable to decode it into IP-level data because of a lack of a special license. After some time it was decided to concentrate on other testing directions because even in case of success all planned attacks analysis will be too complicated because of the lack of access to SMGW administrator software and even the lack of possibilities to run the SMGW software locally on some other device.

For testing the cases 10 and 12 a communicate with the SMGW in the LMN network is required. For this purpose, we firstly use a RS-485 to UART adapter module based on the MAX485 board. After getting the possibility to analyze SMGW

RS485 data we found that none of the tested by us software (4 python lib's including scapy, 2 c++ lib's, and Wireshark text2pcap) can parse this HDLC data, and was decided to analyze it manually.

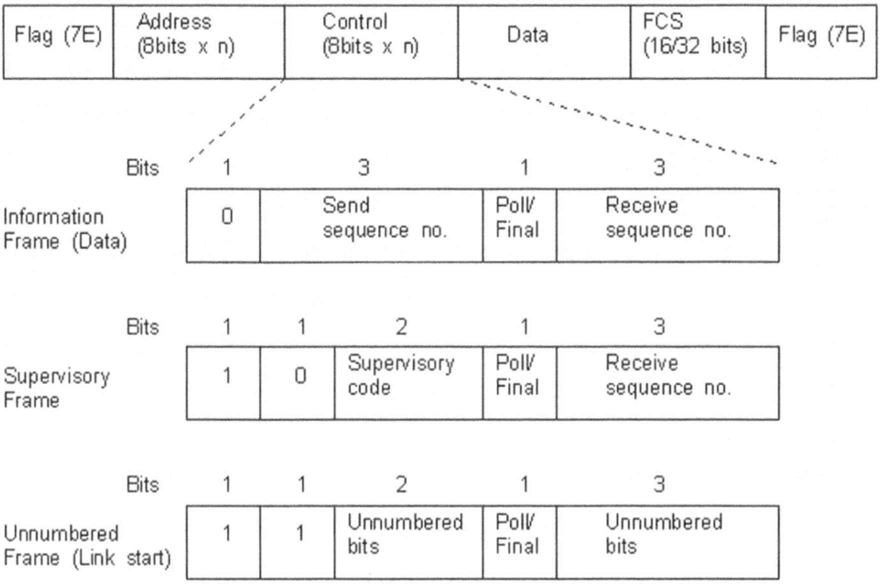

Picture Common HDLC frame structure

As we can see from the picture above—common HDLC implementations include frame start flag 7E field, address field, control field, information field, frame check sequence (FCS), and frame end flag 7E as presented in the picture below. In case of observed data, this sequence differs:

List 3 SMGW CONEXA LMN data

```
7e a0 09 fe03 0203 13 842a 7e
7e a0 2b 0203 b803 131cb6 5c 00
0a01454d4800005ec1f6000000000a01454d4800005ec1f6000000000000 4cec 7e
7e a0 09 b803 0203 53 3a92 7e
7e a0 09 0203 b803 1f 46f1 7e
7e a0 2b fe05 0205 13f8b8 5c 01
0a01454d4800005ec1f6000000000a01454d4800005ec1f6000000000000 7829 7e
7e a0 2b 0205 b805 1356a9 5c 01
0a01454d4800005ec1f6000000000a01454d4800005ec1f6000000000000 7829 7e
7e a0 09 b803 0203 93 3654 7e
7e a0 09 0203 b803 73 2c58 7e
7e a0 09 b803 0203 11 2cf3 7e
7e a0 09 0203 b803 11 3818 7e
7e a0 7a b803 0203 10d92b
160303006a0100006603036140f3c77fed3a4d19f8f9305b67e356222a0e625fdf7c27f00
c6c0bf3f8546000000ac02bc02cc023c02400ff01000033000d0008000604030503060 30
00a000c000a00170018001a001b001c000b00020100000100010300160000001700000
0230000 31d3 7e
7e a0 09 0203 b803 35 1e7f 7e
7e a0 2b fe03 0203 13b2a7 5c
000a01454d4800005ec1f6000000000a01454d4800005ec1f6000000000000 4cec  7e
7e a0 09 b803 0203 11 2cf3 7e
7e  a2  4d  0203  b803  30750d
1603030064020000600303661e9440e22e15479217cdd6c62142d2fXXX
```

Where:

7e—frame start/end flag

a0—some control data OR first part of a frame length field

09—frame length (including this and first byte)

fe03 b803—HDLC address (destination). fe03—is a broadcast address

0203 b803—HDLC address (source). 0203 is probably SMGW address that is always the same in the case of PPC and 0a01454d—some individual smart meter number string that includes EMH (name of smart meter producer) in utf8 form

842a—FCS, frame check sequence

13—some unknown payload parts

160303006aXXX—TLS client_hello payload

1603030064XXX—TLS server_hello payload

As we can see, the flag and address frame fields are the same with a common HDLC, but FCS (frame check sequence) calculates differently. We tried different FCS algorithms with special calculators but were not able to find the correct one. The TLS messages look normal and we can analyze them in hex form, but using some external testing software required additional modification. Because of all described problems was decided to concentrate on HAN attacks for several reasons:

- It is likely that the same implementation of TLS is used across all SMGW networks.

- TCP layer attacks should have the same effect in every network because commonly TCP implementation work on the kernel level.
- It is just more convenient to communicate with SMGW via Ethernet.

SSH Server Testing

The first priority was testing of the SSH server that was found on the SMGW PPC LTE because getting access to it (even in non-root mode) will make testing much more effective because it will allow to understand processes running in the device. It was surprising to discover this SSH server, as the only source of information about it was (https://stg-tud.github.io/sep/projects/2017/eMobilityTeam/site/#technologies). This source also has a few important pieces of information about the testing system:

> As developers, we can access the system via Secure Shell (SSH) with root rights. It is an embedded PTXDist Linux that PPC already uses in the same form for the SMGW. The hardware is an ARMv5 board. Much of the memory (except for /usr/ local/) is persistent. When the system starts, the script /usr/local/bin/custom.sh is called. This can be adapted by the developer in order to generate a certain behavior after a restart. Package management is available with Open Package Management (OPKG) so that the software can be packed into a Debian package and installed.
>
> Due to the low computing power (ARM926EJ-S with ~ 226 BogoMIPS), the limited memory (~ 250 MB main memory, ~ 200 MB free permanent memory) and the hardware-related requirements, the SEI software will consist almost exclusively of C and C++ components.

Manual connection to the PPC SMGW HAN SSH server shows that the server accepts username/password authentication (not only certificate). Authentication log:

List 4 PPC SMGW HAN SSH server manual connection log and common SSH connection log

```
ssh root@192.168.100.100 -p 2222
root@192.168.100.100's password:
Permission denied, please try again.
root@192.168.100.100's password:
Permission denied, please try again.
root@192.168.100.100's password:
root@192.168.100.100: Permission denied (publickey,password).
```

The technician from the company which provides us with SMGW devices told us that by their information SMGW testing firmware is the same as the production one, and they don't know about such difference. It is possible that SSH may be used by service technicians, but the likelihood is low because SSH provides too wide access capabilities and it is difficult to restrict all prohibited functionality. Moreover, "HAF2: Service technician data provision" requires the use of certificate based authentication instead of password based. We assume that the SSH server is presented only in testing firmware, but even in this case, it can be very useful for our testing.

Because the SSH password authentication method is available—one of the methods to get the password is to use the brute force technique. The most effective SSH brute force can be done if at least the list of machine users is known. Fortunately, used on SMGW Dropbear SSH version 2017.75 is vulnerable to CVE-2018-15599. In Dropbear through 2018.76 function recv_msg_userauth_request in svr-auth.c is prone to a user enumeration vulnerability because username validity affects how fields in SSH_MSG_USERAUTH messages are handled. This is an issue similar to CVE-2018-15473.

The first way to exploit this vulnerability is to compare SSH server response time with different usernames. To check this was created test bench—raspberry pi with the same version of Dropbear SSH 2017.75 and was created Paramiko lib based python script. Because the list of users available via SSH in a raspberry pi test bench is known—the testing approach includes SSH username request-response timing comparing different usernames with the same big-length password. On the test bench USB-Ethernet PC connection ping time is an average of 0.57 ms. Via direct PC PCI-Ethernet to SMGW HAN Ethernet average ping time was 0.47 ms. But even with this timing on the test bench it was impossible to distinguish available and absent SSH users, probably because the used python library is too slow for this purpose.

An alternative way to test this vulnerability is to use the Metasploit auxiliary/scanner/ssh/ssh_enumusers module. This Metasploit module includes two options—"Timing Attack" and "Malformed Packet Attack". The timing attack option is close to the previous method and is based on the fact, that some versions of SSH will return a "permission denied" error for an invalid user faster than for a valid user. With the Malformed Packet option, Metasploit sends a malformed (corrupted) SSH_MSG_USERAUTH_REQUEST packet using public key authentication (that must be enabled) to enumerate users.

As in the case of using the python SSH timing script—The Metasploit ssh_enumusers module with the "Timing Attack" option does not lead to any results even in the raspberry pi test bench. But Metasploit Malformed Packet Attack gives results and we got a list of usernames for which SSH authentication is enabled and in the case of the raspberry pi test bench and on PPC SMGW. The received list of users did not allow speed up brute-force or make any assumptions about the operating system or other software running on the SMGW because these user names are present in most Linux-based systems.

The next brute-force step is to try to find a password for found usernames. There exist a lot of instruments for SSH credentials brute force attacks. The most popular are Metasploit ssh_login module with PASS_FILE option, Nmap NSE ssh-brute script, Patator, Medusa, and Hydra. Metasploit and Nmap options are the slowest in comparison with the other software because they can do brute force only with one thread. Hydra has the advantages to include the possibility to use several threads to improve the brute force speed and the possibility to continue from the point of word list after a brute force stop. Developers recommend using only 4 threads with low-performance devices to not cause a device DOS. With 4 threads speed of SSH brute force in PPC SMGW is only about 150 tries/min. Hydra PPC SMGW SSH brute force tests show that it is also possible to use 5 threads. When was used 6 threads or more we got server responses "Connection reset by peer" on some of the requests.

At the moment, no password was found for any of the users. Perhaps the password is not present in the dictionary. Due to the low speed of password brute force, likely, it will not be possible to find the correct password until the end of the project. There are number of known vulnerabilities in Dropbear 2017.75, but none of them allows to bypass authentication or leads to remote core execution. Password sniffing is not important in our case because we do not have a client who knows the password to intercept it.

Web Servers Testing

Conexa SMGW allows to end-user consumption data retrieve only using special software such as TRuDI from PTB. TRuDI (Transparenz-und Displaysoftware) is a manufacturer-independent, standardized visualization solution that meets the requirements of the MsbG (especially §35, §62), the PTB-A50.8, and within the framework of the BSI specifications. TRuDI offers a display function with which the measured values that are available in the SMGW are displayed for the end consumer. In addition, a so-called transparency function is available. As part of this functional feature, the end consumer is able to use the software to locally understand tariff calculations that have been carried out on the basis of the measured values of the SMGW in the supplier's system landscape and thus to check his invoice.

For authentication, TRuDI allows using certificates (HKS1) or Username-Password pair (HKS2). Also, establishing a TRuDI connection requires a SMGW identification number, IP address, and port number. An identification number is required to determine the SMGW model and version in order to select the correct API (the location of this information may vary depending on the company that manufactures a SMGW).

On the HAN TLS port, Conexa SMGW provides a web page on the address https://192.168.0.1/smgw/cust/con-9910634000006-1273.sm/ where cust—probably means customer, con-9910634000006-1273—is user number and, sm—is a part from SMGW DNS. This web page includes only two tabs—the home page and the self-test. Access to this web page is protected via the same username-password as in TRuDI. We were not able to determine which framework or programming language was used via exploring web page code and based on other information. The home page just shows information about the SMGW such as the firmware version, SMGW ID, status, and system time. The self-test page includes only one button to run the self-test. After finishing the SMGW self-test for some reason the are results commonly not showing. Multiple calling self-test functions in a short time range (e.g., by python GET script) does not lead to any harm such as denial of service (DOS). Because the Conexa SMGW web server does not have any fields to enter some data—common SQL injections were tested only in the login and password fields.

PPC SMGW web server includes more possibilities for penetration testing compared with Conexa SMGW. It is accessible on URL /cgi-bin/hanservice.cgi, which means that CGI or common gateway interface technology was used on top of the Lighttpd service.

Common Gateway Interface (CGI) is a web server interface that enables standardized data exchange between external applications and servers. It is one of the first Internet interface technologies which is still widely used nowadays. In the case of using CGI, HTML pages are dynamically generated after the user requests a website and as an advantage, this means that HTML pages do not need to be stored on a server. This means that a CGI script cannot execute directly from a browser. To use a CGI script, it must be located on the same machine where a server is located.

When a user requests a server, this data is first processed by a CGI script. Then, the data is transferred via a standardized CGI interface, which can display the newly generated information in HTML form. CGI scripts are usually stored in a special folder on the web server. A CGI script can be implemented by a wide variety of programming languages. The Common Gateway Interface ensures that the web server and script can communicate with each other, independently of the used programming language. This programming language independency is based on a web server and CGI script interaction—because commonly it uses standard console output.

CGI also has some disadvantages, such as increased performance requirements of the webserver. Every time the CGI application is accessed, a new process is spawned with all the resulting overheads. If the application is buggy, it can loop for example. The browser will terminate the connection when the timeout expires, but the server-side process will continue until the administrator forcibly removes it.

Another problem is even with a low server load the response time of the CGI application is sometimes quite long because a script has to be restarted for each new input. Especially for websites with high traffic, it can become a problem that servers often only support a certain number of CGI applications and further incoming requests are then put into a queue or rejected—but for SMGW is not a big problem because conception assumes a low number of users. But even without a DOS attack, the PPC SMGW web server works slowly, from 2 to 10 s to open just 1 page. Commonly it required 5 s before getting the login form and about 5 more seconds to open a web page. Taking all the above disadvantages into account, we assume that PPC uses CGI—because it is a widespread and free standard.

SMGWs web server testing is not so different from common web server testing. The main difference is that the number of fields and points which possible to test is minimal. PPC SMGW web server includes more fields and forms to test compared with Conexa SMGW. PPC SMGW web server includes logs, counter, evaluation profile, communication profile, self-test, and software version pages. Logs and counter pages include fields that allow entering dates, and the evaluation profile page includes a drop-down menu. All these fields were tested for common SQL injections and successfully bypass this testing. Testing both SMGWs via Greenbone OpenVAS, Nikto, Burp suite (incl. spider and burp intruder testing), and OWASP ZAP does not allow us to find any vulnerabilities.

Also, because zenmap was able to fingerprint that PPC SMGW use Lighttpd, but was not able to fingerprint the exact version, was tested a few exploits for such CVEs as 2010-0295, 2011-4362, and FastCGI Header Overflow. None of the tested exploits were successful.

TLS Servers Testing

One of SMGWs penetration testing directions is TLS server testing. First, parameters such as TLS server version, cipher suites, or extensions supported by the SMGW are checked. For this purpose, it is possible to use instruments such as sslscan or sslyze. Both SMGWs support only TLS version 1.2 and use ECDHE-ECDSA-AES256-GCM-SHA384, ECDHE-ECDSA-AES128-GCM-SHA256, ECDHE-ECDSA-AES256-SHA384, ECDHE-ECDSA-AES128-SHA256 cipher suites as required by BSI-TR-03109-1. The difference between SMGW's TLS servers—server key exchange group in the case of Conexa SMGW uses 192 bits secp384r1 (NIST P-384) and SSL Certificate ECC Curve secp384r1. PPC SMGW uses 128 bits secp256r1 (NIST P-256) and SSL Certificate ECC Curve prime256v1.

Next, common TLS protocol vulnerabilities were checked. These vulnerabilities can lead to DoS or acceptance by TLS server parameters, which should not be accepted in normal circumstances (for example some weak cipher suites). This testing in HAN is important because as it was mentioned before, LMN and WAN SMGW networks probably use the same TLS implementation. For checking common TLS server vulnerabilities, Java-based TLS-attacker and tlsfuzzer software was used.

TLS-attacker was used to check the following list of common TLS vulnerabilities: Bleichenbacher attack, PSK bruteforcer, invalid curves, heartbleed, lucky13, padding oracle, tls poodle, CVE-2016-2107, early ccs, early finished, and Drown. All tests from this list were successfully bypassed by both SMGW's.

Also, old versions of TLS-Attacker (versions below or equal to 1.2) support TLS server fuzzing instrument. Fuzzing is a special technique for the automated detection of code errors. The fuzzer software sends the known incorrect (invalid, unexpected, or randomized) data to software under test and then analyses software response or monitoring for exceptions such as crashes, failing built-in code assertions, or memory leaks. Effective fuzzing software can generate semi-valid input data which can be accepted by the software parsing part and lead to some unexpected behavior. TLS-Attacker allows the execution of TLS communication with a required number of variable random modifications (mutations) or specially constructed invalid messages. As an example, during integers mutation TLS-Attacker can apply the following list of modifications—the original integer value can be XORed with random bits, shifted left or right, and increased or decreased by a random value. In addition, specific values can be returned based on a dictionary consisting of a zero value and values causing overflows in specific number representations. Similar strategies are employed by modification of further numeric data types. Byte arrays are modified by applying additional strategies. TLS-Attacker automatically generates modifications that duplicate arrays, remove or insert specific bytes, or shuffle the given byte array. The design of modifiable variables allows TLS-Attacker to make a chain of generated modifications.

To detect buffer boundary violations, integer overflows, or other memory corruptions, the runtime behavior of the TLS library has to be observed. For this purpose, the authors use AddressSanitizer (ASan). ASan is a memory error detector that can be enabled during compilation in recent versions of LLVM or GCC compilers. It is typically used while fuzzing C and C++ applications. If a fuzzer finds a memory error in an application compiled with ASan, the application crashes, prints an error message, and exits with a non-zero code.

In the case of SMGW testing, this method does not work, since we do not have access to the TLS server source code or SMGW operating system. Hence, we should produce tests using the "black-box" testing method. For TLS applications, unlike C++ programming languages (like Java) and "black-box" testing cases, TLS-attacker produces a method to analyze the protocol flows with a TLS context analyzer. The TLS context analyzer checks whether a TLS protocol messages flow has been executed correctly contains an invalid protocol flow with an additional protocol message, or a message in a valid protocol flow is modified by a specific modification. In case of a runtime error or an invalid protocol flow, TLS-Attacker stores the protocol flow in an XML file format. This file can later be used for future analysis and TLS messages workflow repeating.

The TLS-attacker fuzzing process occurs in three phases. The starting point for each phase is a set of known TLS protocol flows that includes correct TLS protocol flows and several invalid TLS protocol flows (by default project has 19 workflows). At the beginning of the fuzzing process, TLS-Attacker attempts to execute these protocol flows and stores correctly executed, complete protocol flows for further executions in the next phases.

During the use of TLS-attacker software against Conexa SMGW we observe unusual behavior—SMGW stops responding even at standard TRuDI connection with correct credentials. Also, during the Zenmap scan at the first full cycle of the intense scan, all TCP ports finish normally. But after the second run of intense scans, the SMGW HAN TLS server stops responding. Even after 48 h, the TLS server not came back to working mode. Only device reboot helps to get SMGW back to normal condition.

Zenmap shows that the TLS server port is open and still working normally, but the port state changes to tcpwrapped.

List 5 CONEXA SMGW HAN zenmap analyze

```
Not shown: 999 filtered ports
PORT STATE SERVICE VERSION
443/tcp open tcpwrapped
```

TCP Wrapper is a client-side software solution for Linux/BSD machines that provides firewall features. It monitors all machine incoming packets and if an external node attempts to connect, the software checks if this node is authorized or not based on various criteria that are possible to specify. TCP Wrapper was originally written to protect TCP and UDP-accepting services, but it is also possible to use it to filter certain ICMP packets. When Nmap labels something tcpwrapped, it means that the behavior of the port is consistent with one that is protected by TCP Wrapper. Specifically, it means that a full TCP handshake was completed, but the remote host closed the connection without receiving any data—flag 0x014 (RST, ACK) in Wireshark. It is important to note that TCP Wrapper protects programs (not ports). This means that a valid (not false-positive) tcpwrapped response indicates a real network service is available, but the client is not on a server allowed list of hosts. When a very large number of ports are shown as tcpwrapped, it is unlikely that they represent real services, so the behavior probably means something else like a load balancer or firewall is intercepting the connection requests.

To find parameters that lead Conexa SMGW TLS server to TCP-wrapped condition was created a python script that allows sending TRuDI-like ClientHello TLS messages with some delay pattern. This test shows that for the Conexa SMGW HAN TLS server matters only a number of requests (about 25) and the delay between these requests is not mattered (tested even with 20 min delay). Also, if TLS workflow includes a full TLS connection messages cycle, which means:

1. ClientHello
2. ServerHello, ServerCertificate
3. ServerKeyExchange, ClientCertificateRequest, ServerHelloDone
4. ClientCertificate (empty, contain only Handshake Type: Certificate and Length fields)
5. ClientKey Exchange
6. ClientChange Cipher Spec
7. ClientEncrypted Handshake Message
8. ServerChangeCipherSpec, ServerEncryptedHandshakeMessage.

In this case, the Conexa SMGW HAN TLS server does not stop responding even after 600 full TLS connection messages cycles with 5 s delay. Based on the described information we could conclude that the Conexa SMGW HAN TLS server has additional protection, which is activated on the following conditions:

- If the ClientHello TLS message contains some parameters that do not fit the TLS server settings. The TLS server will not even send a ServerHello message as a response to the ClientHello message.
- Even if the ClientHello parameters are correct, when the TLS connection workflow is not full, the SMGW TLS server will be blocked after 25 incorrect connection attempts.
- Mixing connections with correct and incorrect parameters are not helping. If the counter of incorrect connection reaches 25, the SMGW HAN TLS server will be blocked.
- Changing of Client IP/MAC address did not help to bypass TCPwrapped protection. To check this, an attempt to connect to the SMGW via TRuDI from another machine (with different IP and mac address) in the same local network was done. But theoretically, TCPwrapper software should accept data from an unblocked IP address.

This additional protection was a problem for testing software such as fuzzer in the case of Conexa SMGW because making reboots in the required time gap is not a common task. In the case of PPC SMGW, it also has a similar protection, but SMGW is back to normal condition without a reboot after some timeout (about 300 s), which allows us to simply increase all timeout settings in testing software to use it. TLS-attacker fuzzer does not allow us to find some vulnerabilities in the case of PPC SMGW, and because of the reboot requirement, we were not able to use it correctly against Conexa SMGW.

Another TLS server testing software that was used is TLSfuzzer. TLSfuzzer is not a fuzzing software in common sense—because it is a python library for convenient interaction with TLS protocol messages, which repository includes a list of tlsfuzzer lib-based testing scripts (at current moment 145). Some of these testing scripts such as test-record-size-limit test or test-cve-2016-2107, and 5 scripts for fuzzing TLS plaintext, MAC, padding, ciphertext, and finished messages. The main disadvantage of tlsfuzzer fuzzing scripts—is that scripts have a finite (and comparatively small) set of tests. For example, the padding fuzzing script produces only 11 tests. Ciphertext fuzzing script only 40 tests. But because of a large number of testing scripts for TLS and clear source codes makes this tool useful for the project. In the case of tlsfuzzer testing—all tlsfuzzer tests were successfully passed or show indefinite conditions for both SMGWs.

Because of the SMGW's TCPwrapper protection problem for Conexa SMGW and the small number of reference workflows in the TLS-fuzzer software, it was decided to make our own SMGW fuzzing software based on TLS Response-Guided Differential Fuzzing approach which was presented in the paper "Maximizing and Leveraging Behavioral Discrepancies in TLS Implementations using Response-Guided Differential Fuzzing" (Walz and Sikora 2018) from Andreas Walz and Axel Sikora.

Differential fuzzing is a technique used to find vulnerabilities in software by sending inputs to multiple implementations of the same software and comparing the responses. The assumption behind differential fuzzing is that if two different implementations of the same software behave differently when given the same input, there may be a vulnerability in one of the implementations. This technique can be particularly useful for testing implementations of protocols like TLS, where it can be difficult to determine whether an implementation is correct or not. To use differential fuzzing, it is important to first set up the environment with multiple implementations of the software which will be tested. You can then use a fuzzing tool to generate a large number of random inputs and send them to each of the implementations. Next possible to compare the responses from each implementation. If some of the response is not the same, probability of vulnerability in the place of TLS implementation which was activated by this request is much higher.

It is important to note that differential fuzzing is not a foolproof method and should be used in conjunction with other testing techniques to ensure the most thorough testing possible. However, as shown on paper, this approach allows us to get better testing and TLS server source code "covering" compared to TLS-attacker or NEZHA (Walz and Sikora 2020). Also, for Conexa SMGW we added electricity smart plug software control support for our fuzzing software which allowed us to automatize fuzzing process.

To determine if anything goes wrong during fuzzing without access to the device shell, we have checked the following list of cases:

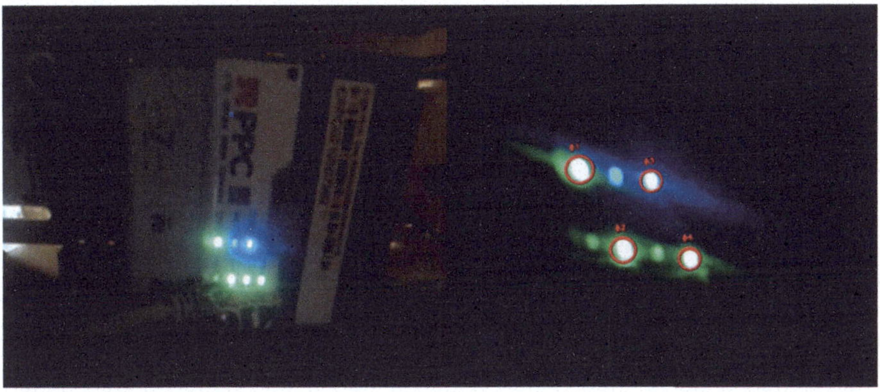

Fig. 9 Watch-dog software example

1. The device Web server does not respond.
2. The device TLS server acts differently in comparison with other TLS servers.
3. The device TLS server does not respond.
4. The device does not respond on the TCP layer.
5. Physical—the interface is down/non-standard LEDs blinking.

Devices LEDs blinking are described in the user manual. We use python OpenCV script to automatize led blinking analysis (Fig. 9). It takes a picture of the device as its input and applies few filters. It then analyses the resulting image (by knowing the comparative LED's position), and returns a list where every element describes the current LED state. But even after that we still don't have a full understanding of what is going on inside the device because we have no shell, and the testing speed was comparatively low, as the tested devices were not so powerful.

As a result of testing, using our fuzzing software against SMGW at the current stage—we were able to observe different behavior of SMGW's HAN TLS servers compared with reference TLS implementations but did not find a way to use it to make some harm.

Another possible direction of TLS server testing is using some TLS certificate vulnerabilities such as CVE-2019-3829 which allows making memory corruption (double free) during the certificate verification (NVD—CVE-2019-3829). Any client or server application that verifies X.509 certificates with GnuTLS 3.5.8 or later should be affected. There are even a number of publicly available exploits for such vulnerabilities (as an example, https://www.exploit-db.com/exploits/46626) but without the knowledge of TLS server version and implementation, this testing is problematic due to a big number of variations.

Because of that, our next work was focused on developing a TLS server fingerprinting software which is based on the same "Response-Guided Differential Fuzzing" approach. This software requires SMGW's TLS implementation and version. Knowing TLS server implementation and version will allow to run tests on a local PC, which is important to increase testing speed and allows to use of address sanitizer for software memory analysis. Unfortunately, during work on this software, we faced several problems that would require more than the expected time to get resolved. Hence, our TLS fingerprinting approach will be described more precisely later in a different paper.

Other Testing Directions

Another possible way to influence Conexa SMGW is to use open the Socks5 port, which is required to implement HKS3—A transparent communication channel initiated by CLS. In the case of the HKS3, the connection must be established using SOCKSv5 [RFC1928] and "TLS for SOCKSv5" [DRAFT-IETF-AFT-SOCKS-SSL-00], which requires establishing of TLS connection before creating SOCKS5 proxy channel. Different Socks5 server authentication methods were tested—and it was observed that Conexa SMGW Socks5 accepts only authentication method 0x86 described in the draft RFC Secure Sockets Layer for SOCKS Version 5. Multiple connection attempts to the Socks5 server do not lead to Conexa SMGW denial of service (which was tested using a python script).

Because of the small number of running on the device services probably the most potentially effective SMGW testing direction is TCP-layer attacks. In most Linux-based systems, the TCP/IP stack is integrated into the kernel, which means that even if some tested port is in a close state or it is implementing a "Wake-Up" service (because, by our knowledge, it is implemented as a listening port on some host system as it shown on page 15 in Detken et al. 2016), it could be vulnerable to TCP-layer attacks. One more advantage is that if the HAN interface is vulnerable to a TCP-layer attack, it is highly probable that LMN and WAN will also be vulnerable to this attack (because of TCP stack kernel integration).

An example of a TCP-Layer attack is the CVE-2019-11815—Remote code execution in Linux kernel TCP/IP implementation. The vulnerability exists due to a race condition that leads to a use-after-free error when TCP packets in rds_tcp_kill_sock() function in net/rds/tcp.c. A remote unauthenticated attacker can specially craft TCP packets to the affected system, trigger a use-after-free error, and execute arbitrary code on the target system. Successful exploitation of the vulnerability may allow an attacker to compromise a vulnerable system (https://www.cybersecurity-help.cz/vdb/SB2019051302). By the data from source (NVD—CVE-2019-11815) vulnerable should be all Linux systems with kernel version lower than 5.0.8, which is true for tested SMGWs by the zenmap data (which can be wrong).

Another example of TCP-layer attacks is Remote DoS in TCP/IP implementation in Linux kernel (https://www.cybersecurity-help.cz/vdb/SB2019061702) (CVE-2019-11477, CVE-2019-11478, and CVE-2019-11479). In the TCP protocol, each segment has a sequence number that indicates the position of the data contained in the packet within the overall stream of data being transmitted. When two systems establish a TCP connection, they exchange initial sequence numbers (ISNs) and use them to initialize a sequence number counter. The sequence number of each subsequent packet is then determined by adding the length of the data in the packet to the previous sequence number. If an attacker sends packets with sequence numbers that are out of order, it can cause problems for the recipient. For example, if an attacker sends segments 3, 4, and 5 before segment 2, the recipient will not be able to properly reassemble the data stream.

To help mitigate this problem, the TCP protocol includes a feature called Selective Acknowledgment (SACK). When SACK is enabled, the recipient of a TCP connection can send a special type of acknowledgment (ACK) packet that includes a list of the sequence numbers of the packets it has received. This allows the sender to identify which packets have been received and which ones have been lost or are out of order, and to retransmit the missing packets as needed.

If an attacker sends segments 3, 4, and 5 before sending segment 2, and SACK is enabled on the recipients end, the recipient will buffer segments 3, 4, and 5 and send duplicate ACKs for them when they arrive. However, it will not be able to fully reassemble the data stream until it receives segment 2, so it may lead to a buffer overflow.

There are no publicly available exploits to test these CVEs, but from the point of view of security, it is good. In the public access, only one exploit for CVE-2019-11477 was found (GitHub—sasqwatch/cve-2019-11477-poc), and even that was not working properly on the local system (with the required exploit parameters).

7 Recommendations

The BSI SMGW has a great protection level because even in case of a small number of points of entrance and great elaboration (which is shown by the "Wake-Up" service), it has additional security mechanisms such as "TCP Wrapper" which makes penetration testing even more difficult. Both SMGW devices have minimalistic web servers. Forms/APIs which allow user input was protected (escaped) against tested injections. PPC SMGW SSH server is vulnerable to the information-disclosure vulnerability, but to our knowledge, this SSH service is preserved only in testing SMGW firmware. As we can see testing of such devices requires not an "in-wide" approach as in the common penetration testing but an "in-depth" approach, because the number of possibilities to test are low. Such tasks require much more effort and knowledge from the tester.

Based on materials from this chapter it is possible to give some recommendations because even taking into account all the reviewed systems and their differences, the most likely successful attack vectors are the same:

- A good practice is to apply security control at all stages of the development—design, implementation, product decommissioning, and maintenance.
- Always make security updates/patching in time. Currently, there is no information about any vulnerabilities in the reviewed systems parts. But this does not mean that they will not appear in the future. For example, the source (Marcellin 2018) mentioned that Enedis hubs use the Java programming language. The author pointed out that Java has suffered for years from chronic vulnerabilities, which Java publisher Oracle—patches month after month. In December 2021 has been

discovered a serious vulnerability in a very widespread Java library for logging service Log4J (CVE-2021-44228), which theoretically can exist in some Enedis software.

- Make hardware upgrades in time. Theoretically old Linky meters can be vulnerable because in document PLC profile specifications (ERDF-CPT-Linky-SPEC-FONC-CPL) from the year 2009 (but it is the newest founded version) refers to Green Book Cosem DLMS UA 1000-2:2008 edition 7, which does not contain such important DLMS/COSEM features as asymmetric cryptography that allow producing digital signatures and a secure key establishment and management.
- It is a good practice to use common protocols such as TLS or DLMS/COSEM and the most well-known implementations of these protocols to secure a system. Because these implementations have been tested many times and are guaranteed to provide a high security level if they are used with the correct settings. TLS and DLMS/COSEM protocols offer approximately same level of security in smart metering solutions. Both use PKI, modern security primitives, and cipher suites. DLMS/COSEM is comparatively more lightweight. The advantage of TLS is that this protocol is more widely researched for potential vulnerabilities because it is more widespread (for example, in banking and shopping business areas where secure Internet connections are very important).
- Users should be entrusted to solve as few security related issues as possible. A professional trained certified operators such as SMGW administrators in the BSI SMGW system are less likely to make some wrong settings which could lead to vulnerability. In conjunction with wide logging and system analysis, suspicious activity can be quickly identified. It is also easier to pinpoint who acted, when, which actions were performed, and the results of this activity.
- It is important to supervise employees legitimate access to systems. To protect against the malicious activity of authorized employees, it is necessary to implement strong auditing, reporting processes and capabilities that can capture user activity.
- Organize staff cyber security/social engineering training more frequently. In most cases, hackers intrude inside a secure system using attacks involving a human factor (Positive technologies 2018), like in the case of the Ukrainian power station in December 2015, where hackers used email with spear-phishing Microsoft office document (CVE-2014-4114) containing malicious macros to turn off power substations (Macola 2020).
- Check HES/SMGW admin/EMP networks perimeter or use a special intrusion detection system periodically. This is potentially the most interesting part for hackers because they have the highest probability of making profit in the case of a successful hack. They also have potentially more system entry points here.
- Tamper resistance is very important to protect the system from an internal attacker. The mechanisms such as encrypted communication, signed and verified firmware, disabled debug communications interface (such as JTAG), encrypted flash memory, configurable locked optical ports, meter tamper detection, backhaul protection, and other physical and system-level security features should be used.

- Devices should have as few services/interfaces as possible because it will decrease the number of possible entry points. If devices need to use a web server, it is important to try to reduce the number of forms/APIs that allow user input into the system. Those forms/APIs that will be present on the web server must be carefully escaped (most important on the server side) in order to eliminate the possibility to perform SQL/XSS injection. In the case of SMGW penetration testing, these were the main factors that made testing difficult.
- Decrease information about names and versions of services running on the device. As in the case of SMGW penetration testing—getting information about services allows a hacker to switch from a "black-box" type of testing to testing a service on the local machine—which significantly speeds up the testing process and the probability of vulnerability finding.
- If a device like in the case with SMGW HAN is able to connect to the home internet router—check the home internet router security (firewall should be enabled, and all accessible from WAN services should be sufficiently protected).

References

Agence nationale de la sécurité des systèmes d'information. A word from the Director-General. https://www.ssi.gouv.fr/en/mission/word-from-director-general/. Accessed 5 Dec 2022

Agence nationale de la sécurité des systèmes d'information Certification CSPN. https://www.ssi.gouv.fr/administration/produits-certifies/cspn/. Accessed 5 Dec 2022

ANSSI (2014) Detailed measures cybersecurity for industrial control systems

Bartock M, Cichonski J, Franklin J. LTE security—how good is it?

Batra N, Kelly J, Parson O, Dutta H, Knottenbelt W, Rogers A, Singh A, Srivastava M (2014) NILMTK: an open source toolkit for non-intrusive load monitoring. In: E-energy 2014—proceedings of the 5th ACM international conference on future energy systems. Association for Computing Machinery, pp 265–276

BSI (2017) BSI TR-03109-4—Smart Metering PKI—Public Key Infrastruktur für Smart Meter Gateways

BSI Smart Meter Gateway. https://www.bsi.bund.de/DE/Themen/Unternehmen-und-Organisationen/Standards-und-Zertifizierung/Smart-metering/Smart-Meter-Gateway/smart-meter-gateway_node.html. Accessed 6 Dec 2022

BSI Smart Meter: Stellungnahme des BSI zum Eilbeschluss des Oberverwaltungsgerichts Münster. https://www.bsi.bund.de/DE/Service-Navi/Presse/Alle-Meldungen-News/Meldungen/Stellungnahme-OVG-Muenster-Smart-Meter_080321.html. Accessed 6 Dec 2022

BSI Smart Metering Systems. https://www.bsi.bund.de/DE/Themen/Unternehmen-und-Organisationen/Standards-und-Zertifizierung/Smart-metering/smart-metering_node.html. Accessed 6 Dec 2022

Carracedo G (2019) Smart meters in Spain—telemanagement system | Tarlogic. https://www.tarlogic.com/blog/smart-meters-spanish-scenario-telemanagement/. Accessed 4 Dec 2022

CEN/CLC/ETSI/TR 50572 (2011) Functional reference architecture for communications in smart metering systems

Chauvenet C (2016) G3-PLC, the standard of the LINKY roll-out and beyond

CNIL (2014) Pack de conformité sur les compteurs communicant

Compteur Linky Obligatoire : comment refuser la pose et date limite. https://www.fournisseurs-electricite.com/guides/compteur/linky/refuser. Accessed 5 Dec 2022

CRE (2014) Délibération de la Commission de régulation de l'énergie du 12 juin 2014 portant recommandations sur le développement des réseaux électriques intelligents en basse tension
CRE (2016) Délibération de la Commission de régulation de l'énergie du 3 mars 2016 portant décision sur la tarification des prestations annexes réalisées à titre exclusif par les gestionnaires de réseaux de distribution d'électricité
Detken K-O, Jahnke M, Humann M (2016) Integritätsmessung von Smart Meter Gateways
DLMS/COSEM protocol security evaluation. https://research.tue.nl/en/studentTheses/dlms-cosem-protocol-security-evaluation. Accessed 5 Dec 2022
Dworkin M (2007) Recommendation for block cipher modes of operation: Galois/Counter Mode (GCM) and GMAC. https://doi.org/10.6028/NIST.SP.800-38d
EDRF (2009) Linky PLC profile functional specifications—ERDF-CPT-Linky-SPEC-FONC-CPL
enedis API—Enedis Open Data. https://data.enedis.fr/api/v2/console. Accessed 5 Dec 2022
Enedis Documentation de référence. https://www.enedis.fr/documents?term_node_tid_depth%5B106%5D=106. Accessed 5 Dec 2022
enedis.fr Enedis company profile. https://www.enedis.fr/qui-sommes-nous. Accessed 5 Dec 2022
enedis.fr Notice d'utilisation du compteur communicant Linky
EUR-Lex (2012) 2012/148/EU: Commission Recommendation of 9 March 2012 on preparations for the roll-out of smart metering systems—EN. https://eur-lex.europa.eu/legal-content/EN/TXT/?uri=CELEX%3A32012H0148&qid=1650551228719. Accessed 5 Dec 2022
European Commission DG Energy (2019) European smart metering benchmark
French National Assembly and the Senate (1978) ACT 78-17 of 6 January 1978 on Data Processing, Data Files and Individual Liberties
Genest O, Maury V, Huc Y. Technical overview of G3-PLC. Trialog. https://www.trialog.com/en/technical-overview-of-g3-plc/. Accessed 5 Dec 2022
GitHub—sasqwatch/cve-2019-11477-poc. https://github.com/sasqwatch/cve-2019-11477-poc. Accessed 8 Dec 2022
gnutls 3.6.6—"verify_crt()" Use-After-Free—Linux dos Exploit. https://www.exploit-db.com/exploits/46626. Accessed 8 Dec 2022
GRDF.FR GRDF: France's leading natural gas distribution operator. https://www.grdf.fr/english/leading-natural-gas-distribution-operator. Accessed 5 Dec 2022
GRDF.FR Le compteur communicant gaz par GRDF. https://www.grdf.fr/institutionnel/actualite/dossiers/compteur-communicant-gazpar. Accessed 5 Dec 2022
Greveler U, Rhein-Waal H, Glösekötter P, Justus B, Loehr D (2012) Multimedia content identification through smart meter power usage profiles
Gurux for DLMS smart meters. https://www.gurux.fi/front-page. Accessed 5 Dec 2022
Higgins KJ (2014) Smart meter hack shuts off the lights. https://www.darkreading.com/perimeter/smart-meter-hack-shuts-off-the-lights. Accessed 4 Dec 2022
Hoffmann SG, Massink R, Bumiller G (2016) New security features in DLMS/COSEM—a comparison to the smart meter gateway. In: Proceedings of the 2015 IEEE innovative smart grid technologies—Asia, ISGT ASIA 2015. https://doi.org/10.1109/ISGT-ASIA.2015.7387098
i-cube software DLMS security basics. https://icube.ch/Security/security1.html. Accessed 5 Dec 2022
Jöbstl W (2019) Innovations beyond smart metering
Landis+Gyr (2014) Gridstream solution security overview—white paper
Landis+Gyr (2020) Gridstream solution Landis+Gyr HES product description
Landis+Gyr Landis+Gyr DC450. https://www.landisgyr.de/product/landisgyr-concentrateur-dc450-2/. Accessed 5 Dec 2022
Landis+Gyr METAS certificate. https://www.cclab.com/news/landis-gyr-metas-certificate. Accessed 20 Dec 2022
Landis+Gyr White Paper. IDIS interoperability—securing long-term investments with interoperable solutions
Landis+Gyr White Paper. IDIS (Interoperable Device Interface Specification)

Légifrance (2001) Décret n°2001-630 du 16 juillet 2001 relatif à la confidentialité des informations détenues par les gestionnaires de réseaux publics de transport ou de distribution d'électricité, pris pour l'application des articles 16 et 20 de la loi n° 2000-108 du 10 février 2000 relative à la modernisation et au développement du service public de l'électricité
Légifrance (2002) Décret n°2002-535 du 18 avril 2002 relatif à l'évaluation et à la certification de la sécurité offerte par les produits et les systèmes des technologies de l'information
Légifrance (2004) Décret n°2004-183 du 18 février 2004 relatif à la confidentialité des informations détenues par les opérateurs exploitant des ouvrages de transport, de distribution ou de stockage de gaz naturel ou des installations de gaz naturel liquéfié
Légifrance (2005) Loi n° 2005-781 du 13 juillet 2005 de programme fixant les orientations de la politique énergétique
Légifrance (2009) Décret n° 2009-834 du 7 juillet 2009 portant création d'un service à compétence nationale dénommé « Agence nationale de la sécurité des systèmes d'information »
Légifrance (2011) Décret n° 2011-170 du 11 février 2011 modifiant le décret n° 2009-834 du 7 juillet 2009 portant création d'un service à compétence nationale dénommé « Agence nationale de la sécurité des systèmes d'information »
Légifrance (2012a) Arrêté du 4 janvier 2012 pris en application de l'article 4 du décret n° 2010-1022 du 31 août 2010 relatif aux dispositifs de comptage sur les réseaux publics d'électricité
Légifrance (2012b) Délibération 2012-404 du 15 novembre 2012
Légifrance Code de l'énergie
Lüring N, Szameitat D, Hoffmann S, Bumiller G (2018) Analysis of security features in DLMS/COSEM: vulnerabilities and countermeasures. In: 2018 IEEE power and energy society innovative smart grid technologies conference, ISGT 2018, pp 1–5. https://doi.org/10.1109/ISGT.2018.8403340
Macola IG (2020) The five worst cyberattacks against the power industry since 2014. https://www.power-technology.com/analysis/the-five-worst-cyberattacks-against-the-power-industry-since2014. Accessed 5 Dec 2022
Marcellin D (2018) Cybersécurité : Linky, un système IIoT atypique | Alliancy. https://www.alliancy.fr/cybersecurite-linky-un-systeme-iiot-atypique. Accessed 6 Dec 2022
Matoušek P (2017) Analysis of DLMS protocol. Technical Report no. FIT-TR-2017-13
Bundesnetzagentur Messstellenbetriebsgesetz (MsbG). https://www.bundesnetzagentur.de/DE/Beschlusskammern/BK08/BK8_09_MsbG/BK8_MsbG_Basepage.html. Accessed 6 Dec 2022
METAS Datensicherheitsprüfungen durch METAS-Cert. https://www.metas.ch/ds. Accessed 7 Dec 2022
Ministère de l'Écologie (2017) Le déploiement du compteur Linky
Ministère de l'Écologie DDDEDL (2014) Décision du 23 septembre 2014 relative à la généralisation du projet de compteurs communicants en gaz naturel
Molina-Markham A, Shenoy P, Fu K, Cecchet E, Irwin D (2010) Private memoirs of a smart meter. In: BuildSys'10—proceedings of the 2nd ACM workshop on embedded sensing systems for energy-efficiency in buildings, pp 61–66
Murrill BJ, Liu EC, Thompson RM (2012) CRS report for congress smart meter data: privacy and cybersecurity
Nguyen PB. Protection des données et compteur intelligent
NVD-CVE-2019-11815. https://nvd.nist.gov/vuln/detail/CVE-2019-11815. Accessed 8 Dec 2022
Port Scanning Basics | Nmap Network Scanning. https://nmap.org/book/man-port-scanning-basics.html. Accessed 8 Dec 2022
Positive technologies (2018) Positive research 2018. J Inf Secur
PPC Power Plus Communications firmware. https://gwafirmware.ppc-ag.de/. Accessed 8 Dec 2022
Remote code execution in Linux kernel TCP/IP implementation. https://www.cybersecurity-help.cz/vdb/SB2019051302. Accessed 8 Dec 2022
Remote DoS in TCP/IP implementation in Linux kernel. https://www.cybersecurity-help.cz/vdb/SB2019061702. Accessed 8 Dec 2022
RS485 sniffer. http://jheyman.github.io/blog/pages/RS485Sniffer/. Accessed 8 Dec 2022

Rullaud L, Gruber C (2020) Distribution grids in Europe
Siconia® SMARTY IQ-LTE | Sagemcom. https://www.sagemcom.com/V02/de/smart-city/dr-neuhaus/smart-metering/siconiatm-smarty-iq-lte/. Accessed 5 Dec 2022
Sánchez Jiménez M (2011) M/490 EN. Standardization mandate to European Standardisation Organisations (ESOs) to support European Smart Grid deployment
Schweizerische Bundesrat (2008) SR 734.71—Stromversorgungsverordnung (StromVV). Schweizerische Bundesrat
SEP eMobility Team Der Smart Energy Identifier. https://stg-tud.github.io/sep/projects/2017/eMobilityTeam/site/#technologies. Accessed 8 Dec 2022
ShareTechnote 4G/LTE—NAS. https://www.sharetechnote.com/html/Handbook_LTE_EEA.html. Accessed 8 Dec 2022
SMARTER TOGETHER (2019) Report on deployment of Linky smart power meters in the area
BSI Standardisierungsstrategie zur sektorübergreifenden Digitalisierung nach dem Gesetz zur Digitalisierung der Energiewende Roadmap für die Weiterentwicklung der technischen BSI-Standards in Form von Schutzprofilen und Technischen Richtlinien
swissmig (2019) Prüfmethodologie zur Durchführung der Datensicherheitsprüfung für Smart Metering Komponenten in der Schweiz
UFC-Que Choisir (2017) Compteur Linky—Le vrai du faux. https://www.quechoisir.org/action-ufc-que-choisir-compteur-linky-le-vrai-du-faux-n11627/. Accessed 5 Dec 2022
UVEK (2014) Grundlagen der Ausgestaltung einer Einführung intelligenter Messsysteme beim Endverbraucher in der Schweiz Technische Mindestanforderungen und Einführungsmodalitäten
UVEK (2015) Smart Grid Roadmap Schweiz Wege in die Zukunft der Schweizer Elektrizitätsnetze
VSE (2018) Richtlinien für die Datensicherheit von intelligenten Messsystemen
VSE (2019) Intelligente Messsysteme Der Einsatz von intelligenten Messsystemen in der Schweiz
Walz A, Sikora A (2018) Maximizing and leveraging behavioral discrepancies in TLS implementations using response-guided differential fuzzing. In: Proceedings—international Carnahan conference on security technology, Oct 2018. https://doi.org/10.1109/CCST.2018.8585565
Walz A, Sikora A (2020) Exploiting dissent: towards fuzzing-based differential black-box testing of TLS implementations. IEEE Trans Depend Secure Comput 17:278–291. https://doi.org/10.1109/TDSC.2017.2763947
NVD—CVE-2019-3829. https://nvd.nist.gov/vuln/detail/CVE-2019-3829. Accessed 8 Dec 2022

Legal Aspects of the Smart Meter Rollout in Germany, France and Switzerland

Sarah Herrmann and Michael Frey

1 Introduction

Smart meters are a central component of the heat transition, understood as the energy transition of heat, embedded in the goal of all political-administrative levels to achieve the Paris climate targets. This text examines the legal framework for the deployment and rollout of smart meters in Germany, France and Switzerland de lege lata and, on this basis, elaborates proposals for better cross-border smart meter networking.

Basically, the term smart meter refers to digital gas, water or electricity meters ("modern metering equipment") that can also receive and transmit data digitally ("smart meter gateway") and are integrated into a communication network for this purpose (Energie-Handels-Gesellschaft 2022). From a technical point of view, smart meters thus consist of a digital electricity meter and a communication element that enables data transmission.

Legally, smart meters are defined as follows in Germany, France and Switzerland:

The German "Messstellenbetriebsgesetz" (2016) (MsbG), which regulates the operation of metering points and the equipping of grid-based energy supply with modern metering devices and smart metering systems, uses the term "intelligentes Messsystem" for this purpose. According to § 2 No. 7 MsbG, this consists of a modern metering device that is connected to a communications network via a smart meter gateway (§ 2 No. 19 MsbG) and complies with the requirements on data protection, data security and interoperability contained in § 20 and 21 MsbG.

French law defines smart meters as "compteur communicant". However, the term itself is not directly defined by law. The framework conditions for use are derived here from Art. L. 341-4 du Code de l'énergie.

S. Herrmann (✉) · M. Frey
Hochschule Kehl, Kehl, Germany
e-mail: herrmann@hs-kehl.de

D. Ould Abdeslam (ed.), *Smart Meters*, Lecture Notes in Energy 97,
https://doi.org/10.1007/978-3-031-27556-2_6

The Swiss legal system uses the same terminology as German law: smart meters are referred to as "intelligent Messsysteme" (Art. 17a para. 2 StromVG CH and Art. 31e para. 1 StromVV CH), and their introduction is also referred to as smart meter rollout. A smart metering system is defined as a "metering device for recording electrical energy that supports bidirectional data transmission and records the actual flow of energy and its development over time" (Art. 17a para. 1 StromVG CH).

These regulations have their common background and framework in the so-called "Electricity Directive" in the 3rd EU internal energy market package (Directive 2009/72/EC on the internal electricity market), which the aforementioned regulations serve to implement. Cross-border networking of smart meter deployment is initially a technical issue. However, both the technical framework and the (procedural) framework for the rollout are defined normatively.

2 Union Law Background

2.1 Primary Law

In terms of EU law, the first question is the primary legal classification of smart meters and grid-based energy sources. While the smart metering systems themselves can be unproblematically classified as tangible objects and thus as goods within the meaning of art. 28 et seq. TFEU, this is less obvious for electricity. Nevertheless, electricity is also to be classified as goods under EU law (Streinz 2018; ECJ 6/64) (and not as a service, which would then be covered by the freedom to provide services, art. 56 et seq. TFEU). In this respect, the ECJ distinguishes whether the object is imported for its own sake or only assumes an auxiliary function for another commercial service (Classen 2021). The former is the case with electricity, which is subject to the same primary law regime as other energy sources such as gas or heat. In terms of competences, art. 194 TFEU also contains an independent allocation of competences to the EU for the area of energy policy (Streinz 2018). This was also intended to achieve a clear separation of energy policy from environmental policy and to standardize the previously existing "patchwork" of competence standards (Kahl 2009).

2.2 Secondary Law

The secondary legal framework is determined by the regulations from the 3rd internal energy market package already outlined above, consisting of Regulation (EC) No. 713/2009 establishing an Agency for the Cooperation of Energy Regulators, Regulation (EC) No. 714/2009 on conditions for access to the network for cross-border exchanges in electricity, Regulation (EC) No. 715/2009 on conditions for access to the natural gas transmission networks, and the Directives concerning common rules for the internal market in electricity (2009/72/EC), so-called Electricity Directive, and for the internal market in natural gas (2009/73/EC). According to Annex I (2) of the Electricity Directive, member states are obliged to introduce smart metering systems after first carrying out a cost-benefit analysis. The member states had until March 2011 to implement the directives (art. 49 of the Directive).

According to the recitals of the Electricity Directive, Member States should contribute to the modernisation of distribution networks, e.g., through the deployment of smart grids, in order to benefit decentralised energy production and energy efficiency (European Parliament Council 2009). However, Member States may make the roll-out of smart metering systems conditional on an economic assessment, the so-called cost-benefit analysis (European Parliament Council 2009). In doing so, the evaluation may lead to the conclusion that roll-out is only economically rational and cost-effective for certain consumers, which Member States should take into account when introducing them (European Parliament Council 2009). Programmatic sentences, which include the recitals of a directive, do not give rise to any legal obligations on the Member States (Wieser 2011).

In the main body of the Directive, Member States or competent regulatory authorities should also strongly recommend that electricity undertakings promote energy efficiency or optimise electricity consumption. This can be done, for example, through the development of novel pricing models, energy management services or, where appropriate, the deployment of smart metering systems or smart grids (European Parliament Council 2009). Due to the wording chosen by the European legislator to 'recommend' and 'expressly', the Member States are not required to make smart meters mandatory in their country (Wieser 2011).

However, in the Annex to the Directive, Member States will have 80% of final consumers equipped with smart metering systems and a timetable for action by 2020. However, these European objectives may be subject to a cost-benefit analysis (European Parliament Council 2009).

The main coordinating body at European level is the Smart Grids Task Force. It advises the European Commission on policy and regulatory directions and was tasked with coordinating the first steps to implement smart grids under the Third Energy Package (European Commission 2022) In March 2012, the European Commission adopted Recommendation 2012/148/EU (European Commission 2012) on preparations for the deployment of smart meters in the Member States, highlighting in particular recommendations on cost-benefit analysis and the usefulness of data protection by design (Säcker and Zwanziger in Berliner Kommentar zum Energierecht) A recommendation is a non-binding instrument of EU (Ruffert in AEUV Art. 288, recital 97) action and has no legal force.

In addition to the Electricity Directive, Art. 9 (1) of the Energy Efficiency Directive (European Parliament/Council: Directive 2012/27/EU) contains requirements for the introduction of individual meters that accurately reflect the end customer's actual energy consumption and provide information on the actual time of use. Unlike the requirements of the Electricity Directive, however, these meters do not have to be intelligent (capable of communication) (this follows in particular from Art. 9(2) of the Energy Efficiency Directive).

Furthermore, the regulations of the General Data Protection Regulation (Regulation (EU) 2016/679) (GDPR) are of central importance at the secondary law level, the regulations of which can claim priority of application in the scope of the Regulation, in particular with regard to personal data.

3 The German Way—Implementation of the Smart Meter Rollout in German Law

3.1 General Information on the Smart Meter Rollout in the MsbG

The German legislature implemented the requirements of the European directive in the Act on the Digitization of the Energy Transition "Gesetz zur Digitalisierung der Energiewende" (Gesetz zur Digitalisierung der Energiewende, 29.08.2016, BGBl I 2016) (GDEW). The centerpiece of this article law was the Metering Point Operation Act "Messstellenbetriebsgesetz" (MsbG). In accordance with the legal basis in § 29 (1) MsbG, the German path provides for mandatory introduction of smart meters in three case groups depending on consumption:

1. for end consumers with an annual electricity consumption of more than 6000 kW h,
2. for end consumers who have agreed a reduced network charge for a controllable consumption device (e.g., heat pump), and
3. for system operators with an installed capacity of more than 7 kW.

About 5 million end consumers are affected by the mandatory installation (Bundesnetzagentur, Bundeskartellamt 2019).

However, the prerequisite for mandatory installation is that it is technically possible in accordance with § 30 MsbG and economically justifiable in accordance with § 31 MsbG. According to § 30 MsbG, technical feasibility requires that at least three companies offer smart metering systems that comply with the standards set out in §§ 19–23 MsbG and that the Bundesamt für Sicherheit in Informationstechnik (BSI) establishes this (§ 30 p. 1 MsbG, so-called "market declaration"). The determination is made as an administrative act in the form of a general ruling [§ 35 p. 2 Verwaltungsverfahrensgesetz (VwVfG)].

With regard to the economic justifiability of the equipment, § 31 MsbG provides for various gross price ceilings for consumers (§ 31 p. 1 MsbG) and installations (§ 31 p. 2 MsbG).

For consumers below 6000 kWh and above 100,000 kWh (Bundesamt für Sicherheit in der Informationstechnik 2022) per year, as well as for operators of systems up to and including 7 kW capacity, installation is optional (§ 29 p. 2 MsbG); it is recommended if it is technically feasible and economically justifiable. The decision on installation is made by the basic metering point operator, who has a so-called "optional installation right"; in addition, the owner can decide on installation in the case of private property, and the landlord in the case of rented property. Tenants do not have a veto right. They must tolerate installation (§ 36 p. 3 MsbG), but must be informed three months in advance with reference to the free choice of a metering point operator (§ 37 p. 2 MsbG).

The option of installing smart metering systems is available for around 39 million end consumers (Bundesnetzagentur/Bundeskartellamt: Monitoringbericht 2019). Metering point operators must equip 10% of the metering points concerned with smart metering systems within three years of the start of the equipment obligation (§ 45 p. 1 no. 1 MsbG). The minimum technical requirements for smart metering systems are defined in § 21 MsbG.

3.2 The Decision of the OVG Münster from 04.03.2021 and the Consequences

In a general ruling dated January 31, 2020, the BSI issued the market declaration pursuant to § 30 p. 1 MsbG. The market declaration results in the installation of smart metering systems becoming mandatory, § 29 p. 1 MsbG. At the same time, deviating metering systems that deviate from the requirements of § 19 p. 2 and p. 3 become inadmissible (Jahn in Stopp für Einbauverpflichtung intelligenter Messsysteme) The general ruling ordered immediate enforcement, which initially eliminates the suspensive effect of any objections [§ 80 (2) No. 4 Verwaltungsgerichtsordnung (VwGO)] (Bundesamt für Sicherheit in der Informationstechnik 2022) Due to an application filed by a competitor to restore the suspensive effect of the objection (§ 80 p. 5 VwGO) against the order of immediate enforcement of the BSI's general ruling of January 31, 2020, the Münster Higher Administrative Court suspended the immediate enforcement of the general ruling by emergency order on March 4, 2021 (Münster 1162). The OVG Münster based its decision firstly on the fact that no three independent companies offer smart metering systems that meet the requirements of § 24 p. 1 MsbG and secondly on the fact that § 30 p. 1 MsbG does not authorize the BSI to restrict the installation of smart metering systems to certain groups of cases (Jahn in Stopp für Einbauverpflichtung intelligenter Messsysteme; Säcker and Zwanziger in Berliner Kommentar zum Energierecht) Due to legal uncertainty, the smart meter was on ice for more than a year. The decision of the OVG Münster led to the reaction of the legislator, who made adjustments to the MsbG. On this basis, the BSI issued a new market declaration on February 8, 2022 in accordance with § 30 p. 1 MsbG (Bundesamt für Sicherheit in der Informationstechnik 2022) On 25 May 2022, the BSI withdrew the contested market availability declaration and adopted a transitional arrangement pursuant to § 19 p. 6 MsbG to ensure the rollout (Haufe 2022)

3.3 Data Protection—Legal Framework of Smart Meters Between DSGVO—BDSG and LDSG as Well as the Sector-Specific Regulations of §§ 55 MsbG

Technical Framework for Data Protection

The Energiewirtschaftsgesetz (EnWG) required not only to respect data protection by means of data protection principles, but also to enshrine data protection in technology (Jandt et al. 2011).

Due to the huge amount of data, smart metering systems require a high level of data protection from a technical point of view. For example, when the device is designed, data protection is already ‘built in’ by the so-called ‘privacy-by-design’ (Deutscher Bundestag: Bundestag-Drucksache 18/7555) solution. The need for corrective programms to respond to security vulnerabilities (Renner 2011) will be minimized and processing operations will be reduced. In order to ensure a ‘privacy-by-design’ approach, the BSI, in consultation with the Bundesnetzagentur (BnetzA) and the Bundesbeauftragter für den Datenschutz und die Informationsfreiheit (BfDI) (Jandt et al. 2011), developed protection profiles (Bundesamt für Sicherheit in der Informationstechnik 2022) pursuant to § 21e(2) EnWG (repealed) in accordance with Common Criteria EAL4+ (Hellmuth and Jakobs in Informiertheit und Datenschutz beim Smart Metering) and technical guidelines (Bundesamt für Sicherheit in der Informationstechnik 2022). These protective profiles and technical guidelines must be complied with in accordance with § 22(2) of the Act, otherwise the equipment may not be operated (Lammel 2022) For the core tasks of the Smart Meter Gateway, which include processing, transmission, storage and erasure, § 22 of the Act lays down high security requirements.

The Structure of Data Protection Law in Relation to the Use of Smart Meters

The data protection framework for the use of smart meters is characterized by the interaction between the regulations at the level of EU law (GDPR) with the “Bundesdatenschutzgesetz” (BDSG) and the different “Landesdatenschutzgesetze” (LDSG) on the one hand and the sector-specific regulations, in particular those of the MsbG (§§ 49 et seq.) on the other hand.

The principle of the primacy of application of EU law applies here between EU law and national data protection regulations (Conrad 2019). Accordingly, national law takes a back seat in the application of the law if and to the extent that an EU law provision contains a corresponding provision. This generally applies to the provisions of the GDPR that are directly applicable pursuant to art. 288 p. 2 TFEU.

However, the GDPR only applies to personal data (art. 2 p. 1 GDPR). If such data is processed, this is initially prohibited unless there is an exceptional case of permission, art. 6 p. 1 GDPR (Wimmer 2020) In favor of national law, the GDPR also contains numerous opening clauses and regulatory provisions. Only in these areas national regulations have to be examined as a matter of priority.

This is also stated in § 1 p. 5 of the BDSG, according to which the BDSG is not to apply in cases where EU law, and in particular the GDPR, "applies directly" (Gusy and Eichenhofer in Beck OK Datenschutzrecht). In this respect, the "two-stage model" is applied here: On the first level, the GDPR applies in principle, the BDSG applies on the second level whenever the GDPR contains an obligatory or optional opening clause (Gusy and Eichenhofer in Beck OK Datenschutzrecht).

In addition, at the level of national law, § 1 p. 2 of the BDSG provides for priority of application in favor of sector-specific special regulations (Gusy and Eichenhofer in Beck OK Datenschutzrecht). Accordingly, the MsbG takes precedence over the BDSG as a lex spezialis, §§ 49 p. 2 MsbG in conjunction with § 1 p. 2 s. 1 BDSG. (Lindermann 2022) However, the fundamental primacy of application of the GDPR also remains in the area-specific data protection law.

This means that the provisions of the GDPR initially apply to personal data, as well as those of the MsbG, whose provisions in the area of personal data must be compatible with those of the GDPR.

Legal Classification of Data Collected by Smart Meters

Smart meters process a large amount of personal data in a largely automated manner and disseminate this data within a communications network (Bretthauer 2017). As end devices, they can be assigned to a connection user by means of an individual identifier (cf. § 2 no. 3 MsbG) and collect extensive information on the user's consumption behavior, for example on personal habits, up to and including the creation of detailed behavioral and personal profiles (Bretthauer 2017). For this reason, the provisions of the GDPR apply primarily to smart meters.

According to art. 4 no. 1 GDPR, personal data is any information relating to an identified or identifiable natural person. This is the case if a person can be identified directly or indirectly by means of association with an identifier such as a name or with location data.

Processing is understood to mean any operation carried out by means of automated procedures, such as the collection, recording, storage as well as the reading and querying of personal data (art. 4 no. 2 GDPR). A smart meter collects a variety of personal data, processes them and transmits them to the energy supplier, for example.

Thus, practically all data collected by the smart meter are to be classified as personal data in the sense of art. 4 no. 1, 2 p. 1 GPDR.

Priority Applicable Provisions of the GDPR

Among the primarily applicable regulations are the principles of art. 5 p. 1 GDPR: Lawfulness, fair processing, transparency (lit. a GDPR); purpose limitation (lit. b); data minimization (lit. c); accuracy (lit. d); storage limitation (lit. e); integrity and confidentiality (lit. f) as well as accountability (art. 5 p. 2 GDPR).

Furthermore, the lawfulness of the personal data processed by the smart meter must first be measured against the justification criteria of the GDPR. This may be consent by the connection user of the smart meter (art. 6 p. 1 a) in conjunction with art. 7 GDPR) or permission (art. 6 p. 1 c) or e) of the GDPR).

Justification on the basis of consent conflicts with the obligation to install smart meters and is therefore ruled out as a justification ground (Bretthauer 2017).

The processing of personal data from smart meters is also lawful under art. 6 p.1c) GDPR if it is necessary for compliance with a legal obligation to which the controller is subject, regardless of whether that legal obligation arises under Union or national law (Bretthauer 2017).

However, art. 6 p. 1 e) of the GDPR does permit data processing by smart meters. According to this regulation, data processing is permitted if it is necessary for the performance of a task that is in the public interest. According to § 29 p. 1 MsbG, metering point operators are obliged to equip a metering point with a smart metering unit if the requirements of p. 1 are met. Art. 6 p. 1 e) GDPR is relevant as a justification even in the absence of an obligation, for example in the cases of § 29 p. 2 MsbG, since the installation of smart meters serves the public interest objective of ensuring a secure energy supply (Bretthauer 2017).

National Special-Law Regulations in the MSbG

Article 6 p. 3 of the GDPR allows Member States to lay down rules on data processing through national law, as has been done in the MsbG.

The special statutory regulations on data protection can be found in §§ 49 et seq. MsbG. § 49 MsbG sets out the general requirements for data processing and lists the entities authorized to process data (including meter operators, network operators and energy suppliers). § 50 MsbG contains the substantive legal basis for the permissibility of data processing (Lindermann 2022) The provision contains a catalog of permissible facts that allow data processing in smart meters, which is generally prohibited, in individual cases and thus directly protects the fundamental right to informational self-determination pursuant to art. 2 p. 1 in conjunction with art. 2 p. 1 Grundgesetz (GG) (Lindermann 2022).

Based on this, §§ 55 et seq. MsbG regulate the permissible scope of data collection as well as the method of measurement value collection. The measurement times and intervals are staggered according to the amount of electricity consumed.

The obligations of the metering point operator, in particular with regard to the processing and transmission of the metering data collected, are described in more detail in §§ 60 et seq. MsbG. § 60 MsbG specifies who receives which data, at what intervals and for what purpose (Lindermann 2022) Whereas before the MsbG came into force, metering data was collected (solely) by the distribution network operators and forwarded by them to the market players (so-called chain communication), § 60 p. 2 MsbG replaces chain communication with star communication (vom Wege et al. 2017). Thus, if smart meters are used, the data collected will in future be transmitted directly to the authorized entities in accordance with § 60 p. 2 MsbG. This means that the network operator will no longer act as a "data hub"; instead, this function will be assumed by the smart meter gateway (Lindermann 2022) The legislator expects efficiency gains and improvements in data protection and data security from the change from chain-based to star-based communication (Deutscher Bundestag: Bundestag-Drucksache 18/7555).

§ 61 MsbG gives the connection user the right to view various information on actual energy consumption and time of use, or to check billing at any time. In addition, according to § 65 no. 1 MsbG, the customer can decide to whom he grants access to his data—beyond his legal obligations to transmit data.

Finally, the use of the data collected and provided by smart meters by network operators, suppliers and balancing group managers is regulated in § 66 et seq. MsbG for each energy market role. §§ 66 et seq. MsbG only refer to the data processing of the metered values received by the market players, but not to the data collection. The regulations on data processing concretize general principles of data protection law, such as the principle of purpose limitation and data minimization (cf. art. 5 no. 1 GDPR) to the scope of application of the MsbG: thus, metered values pursuant to § 66 p. 1 MsbG may only be used for purposes expressly stated in p. 1. § 66 p. 3 MsbG requires the deletion of all personal meter readings as soon as storage is no longer necessary for the performance of the task (principle of storage limitation). Corresponding regulations are contained in § 67 et seq. MsbG for transmission system operators, for balancing group managers and for energy suppliers.

Finally, § 70 MsbG regulates that metering processing or data exchange beyond the regulations specified in §§ 66–69 MsbG is only permissible insofar as no personal data are used or the data subject has consented (art. 6 p. 1 s. 1 lit. a GDPR).

Thus, the provisions of the MsbG ultimately represent only a special legal formulation and adaptation to the technical requirements of general data protection principles. However, the graduation according to consumption volume, which is characteristic of the MsbG and the German approach to smart meter rollout, leads to a not inconsiderable legal complication of the regulatory material.

As is often the case in data protection law, it remains to be seen to what extent the principle of proportionality under the EU Treaty (article 5 p. 1 and p. 4 TEU) and national constitutional law (article 20 p. 3 GG), as well as the fundamental principles of data protection law as set out in article 5 of the GDPR, would not have permitted a less stringent or even unstructured, i.e., simpler and more transparent, structure.

Other Applicable Provisions of National Law

As already described above, direct European law—and thus in particular the GDPR—has priority of application, § 1 p. 5 BDSG. The processing of personal data by federal and state public bodies and by non-public bodies is also covered by the scope of the BDSG (§ 1 p. 1 BDSG).

State authorities and municipalities are also covered by the BDSG in principle (§ 2 p. 2 BDSG), but only insofar as none of the state data protection laws includes more specific provisions or an exception applies (§ 1 p. 2 BDSG).

Unlike the BDSG, non-public entities are not covered by the LDSG.

The term "non-public entities" is defined in § 2 p. 4 BDSG. Accordingly, non-public entities are in principle natural and legal persons, companies and other associations of persons under private law, § 2 p. 4 sentence 1 BDSG. Accordingly, both public and private companies are generally subject to the BDSG, especially those in the energy sector. Municipal enterprises, which are often also active in the energy industry, may also fall within the scope of application of the state data protection laws.

3.4 Interim Conclusion

Due to the German federalism and the staggered and special regulations for smart meters in the MsbG, the regulatory system in German law appears comparatively complex. Although the provisions of the GDPR are primarily applicable to personal data, there are numerous permissions that allow the collection and processing of energy data in the context of the use of smart meters (art. 6 p. 1 c) and e) GDPR). In addition, the regulations of national data protection law apply to the remaining (non-personal) data. Here, most network operators are likely to fall under the scope of application of the BDSG, with the exception of municipal proprietary undertakings.

4 The French Way—Implementation of the Smart Meter Rollout in French Law

4.1 General Information on the French Transposition

French legislation provides for a mandatory one hundred percent conversion to smart electricity meters by 2024 (Code de l'énergie Art. R341-8 par. 3). Although the meter is provided to the customer, it remains the property of the regional authorities (collectivités locales). Consumers cannot object to the installation of the electricity meter. In the event of persistent obstruction of the switch, the customer may be subject to a "special reading" at least once a year, for which a charge will be made (Energie le Lynx 2021).

At the legislative level, Article L341-4 of the Code de l'énergie implements the 2009 directive and states that the electricity supplier "shall make arrangements to offer its customers different prices depending on the time of year or the time of day and to provide incentives to limit their consumption during the periods when the consumption of all consumers is highest." By September 2021, more than 33 million smart meters had been installed in French households (Enedis 2021; Ministère de la transition écologique et solidaire 2021).

The data on total electricity consumption are personal and confidential. They are transmitted encrypted to Enedis remotely using Powerline technology (PLC), which uses the cables of the electricity network for transmission. The monthly data is communicated to the electricity supplier as part of the management of the electricity contract and is subject to end-to-end encryption. Enedis strictly adheres to all recommendations of the Commission Nationale de l'Information et des Libertés (CNIL). According to the network operator, "Linky" enjoys the same level of protection as banks or national defense. In addition, the Agence Nationale de la Sécurité des Systèmes d'Information (ANSSI) regularly conducts security audits to test robustness of the system. In June 2019, ANSSI confirmed the Linky system's ability to withstand cyberattacks (Enedis 2021).

4.2 Legal Framework of the Smart Meter Rollout

At the legal level, article L341-4 Code de l'énergie implements the 2009 European directive. This defines smart meters as a device that allows suppliers to offer their customers different prices depending on the time of year or day, and to create incentives to limit their consumption at times when the consumption of all consumers is highest.

Art. L341-4 p. 2 Code de l'énergie obliges the grid operators to provide the customer with his meter data, warning systems related to the level of individual consumption, as well as comparison elements to local and national averages.

In addition to the general provisions of Art. L341-4, Art. R341-4–R341-8 were added to the Code de l'énergie by the décret Décret n° 2015-1823 du 30 décembre 2015, which defines the scope of application more precisely.

Article R341-8 p. 2 and p. 3 Code de l'énergie prescribes a precise timetable for the French rollout, with the aim of replacing 100% of low-voltage metering equipment up to 36 kV A by 2024. The legislature set an interim target of eighty percent by December 31, 2020, which has been met. According to article R341-8 p. 4 code de l'énergie, 90% of low-voltage installations with power ratings above 36 kilovolt-amperes or high-voltage installations must be retrofitted by December 31, 2024.

According to art. R341-8 p. 2 s. 1 in conjunction with art. L.111-52 no. 1 Code de l'énergie, Enedis is largely (approx. 95%) responsible for converting the old electricity meters to smart meters. In a few regions, such as Alsace, local electricity network operators, so-called Entreprise Locales de Distibution (ELD), carry out the replacement (Électricité de France 2022).

4.3 Data Protection Framework

The framework conditions for processing the data generated and used by the smart meter are regulated in the Code de l'énergie. Certain functions and settings are offered to consumers. Some data is transmitted by default. Others are only collected and used with the customer's consent.

Thus, daily consumption data—the total consumption of the household in one day—is forwarded to the network operator by default. The purpose of this survey is, among other things, to enable customers to view their consumption free of charge (art. R341-4 p. 1 Code de l'énergie).

The collection of finer data—hourly or half-hourly—is not automatic. These detailed data are collected only with the consent of the consumer or selectively, if they are of general interest and necessary for the fulfillment of a public task of the Energy Code (Nationale and de l'Informatique et des Libertés 2022).

Technical Framework Conditions

All consumption data is encrypted. The security measures used comply with the legal requirements (article 4 de l'arrêté du 4 janvier 2012) and the security standards certified by l'Agence nationale de sécurité des systèmes d'information (Anssi).

Encryption keys are set by the installer when the meter is installed and are not replaced later, making hacking more complex. Each smart meter has its own random key. In addition, the smart meter detects physical intrusions (e.g., when the meter cover is opened). In this case, the encryption keys are automatically deleted, making the data inaccessible (Institut national de la consommation 2022) Furthermore, the meters communicate with each other via a concentrator, which has a security box with encryption keys. Finally, the national system is also equipped with an encryption system (Chanvry 2022).

French network operators must comply with the European GDPR. The CNIL issued a decision in 2012 which made recommendations on the collection and processing of personal data from smart meters (Nationale and de l'Informatique et des Libertés 2012). Already in this way, the CNIL has been able to implement the principle of "privacy by design", which came into force in 2016 (Assenblée nationale/sénat 2022) In addition, in 2014 the CNIL, together with the Fédération des Industries Électriques, Électroniques et de Communication (FIEEC), developed a compliance package (pack de conformité) (Nationale and de l'Informatique et des Libertés 2022) as part of a working group, which is intended to support network operators with practical details on compliance with data protection.

The Structure of Data Protection Law in Relation to the Use of Smart Meters

The data protection legal framework in France is also characterized by the interaction of the regulations at the level of EU law (GDPR), it's opening clauses and the general regulations at the national level, such as the loi Informatique et Libertés and the sector-specific regulations, in particular those of the Energy Code (Art. L341-4 and R341-4 et seq. Code de l'énergie) (Sénat 2022).

The decision issued by the CNIL in 2012 and the compliance package published in 2014 are to be classified as soft law due to their recommendatory nature. They are intended to support companies in the implementation and application of European and national regulations.

As already mentioned above in the German chapter, the principle of the primacy of application of EU law, the GDPR, also applies in France.

Legal Classification of the Collected Data

In contrast to the legal classification of the data collected by the German smart meter as personal data, French case law classifies the data collected by the "Linky" smart meter as not being personal data within the meaning of art 4 p. 1 of the GDPR. "Linky" ensures anonymization of the data during its transmission, on the one hand through encryption and on the other hand through the absence of any identifying information. The electricity supplier is only aware of the place of delivery for the creation of the invoice.

It follows that the data subject (art. 4 no. 11 p. 1 GDPR) is not an identified or identifiable natural person within the meaning of art. 4 no. 1 p. 1 GDPR and the network operator does not need to obtain consent from the customer when collecting consumption data (Tribunal de Grande Instance Bordeaux interim order v. 23/04/2019, RG 19/73; 19/75; 19/76; 19/77).

Consent is only required if finer consumption data is collected on an hourly or half-hourly basis for the user's own use or for forwarding to third-party companies. In practice, this does not happen automatically and requires obtaining the user's consent (Tribunal de Grande Instance Toulouse interim order v. 12/03/2019, RG 19/00431).

4.4 Interim Conclusion

Due to French centralism and the few provisions in the Energy Code, the system of rules in French law appears comparatively simple. Although the rules of the GDPR are, in principle, primarily applicable to personal data, French case-law has ruled that the data collected by default is data that is not personal data. The scope of the GDPR will only be opened if the customer wishes to have a narrower interval.

5 The Swiss Way—Implementation of the Smart Meter Rollout in Swiss Law

In Switzerland, the introduction of smart meters has been legally anchored in the Electricity Supply Act "Stromversorgungsgesetz" (art. 17a StromVG) and the Electricity Supply Ordinance "Stromversorgungsverordnung" (StromVV) since January 1, 2018. The data protection aspects are covered by the Swiss Data Protection Act "Datenschutzgesetz Schweiz" (DSchG CH) (art. 17c StromVG).

5.1 Definition and Legal Framework of the Smart Meter Rollout

Art. 17a p. 1 StromVG defines smart meters as an intelligent metering system, i.e., "a metering device for recording electrical energy that supports bidirectional data transmission and records the actual energy flow and its temporal course."

Art. 17a p. 2 StromVG empowers the Federal Council to set requirements for the introduction of such smart metering systems. "In doing so, it shall take into account international standards and recommendations of recognized professional organizations. In particular, it may oblige network operators to arrange for the installation of smart metering systems at all end consumers, generators and storage facilities or at certain groups thereof by a certain date." According to p. 3, it may "determine, taking into account federal legislation on metering, the minimum technical requirements to be met by smart metering systems and the other characteristics, equipment and functionalities they must have, in particular in connection with.

(a) the transmission of measurement data;
(b) the support of tariff systems;
(c) of supporting other services and applications."

The Federal Council has made use of this authorization in the StromVV. Smart meters as intelligent metering systems are defined there in art. 8a StromVV, which defines the technical requirements and, in this respect, corresponds to § 21 MsbG. Art. 8b StromVV ensures that only those smart meters are used that "would be successfully tested to ensure data security." According to art. 8b p. 3 StromVV, the Federal Institute of Metrology is responsible.

In implementing the Energy Strategy 2050, Switzerland is pursuing the goal of replacing 80% of existing electricity meters by 2027 (Schweizerische Eidgenossenschaft/Bundesamt für Energie 2022) and has also stipulated this goal in art. 31e StromVV. Unlike the three-tier system in Germany or the French system, Switzerland has left it up to the grid operator to decide which end consumers and generators he wants to equip with a smart metering system and by when (art. 31e p. 2 StromVV), which represents a considerable flexibilization and simplification. Only "end consumers, when they make use of their entitlement to grid access, [and] generators, when they connect a new generation plant to the electricity grid, must be equipped by the grid operator" (art. 31e p. 2 s. 2 StromVV).

The installation of intelligent control systems according to art. 8c StromVV depends on the consent of the end consumer, generator and storage operator.

5.2 Data Protection Legal Framework in the Data Protection Act

In contrast to the German legal system, the StromVG does not refer to separate technical regulations with regard to data protection, but refers to the general provisions of the Swiss Data Protection Act (art. 17c StromVG).

6 Conclusion

This legal analysis of the smart meter rollout shows how differences in the design of the legal framework can slow down or accelerate the introduction of a technical innovation.

The starting point for these differences is first of all the fact that France and Germany, as Member States of the European Union, have a common normative umbrella in EU primary and secondary law through common Union law. However, secondary law in particular, with the "directive-like" (Wolff and Brink 2019) GDPR, leaves the member states a broad scope through the opening clauses, which ultimately represent the legal reason—in addition to economic policy backgrounds—for the differences between the German and the French path.

With regard to the German path, it should be added that the staggered introduction of smart meters has also led to increased complexity. The standardization of specific data protection regulations in the MsbG as a special law in addition to the general, already federally fragmented data protection regulations also does not contribute to regulatory clarity.

The French approach shows how different the approaches to implementing the directive can be in terms of content, and also how little binding targets were formulated for the member states. In contrast to the German approach, there is no differentiation according to electricity consumption.

In comparison, the Swiss model represents a model for a simple, clear and flexible rollout strategy.

The phenomenon of heterogeneous or incompatible implementation of European framework legislation is a classic problem with directives and is to some extent due to the legal instrument of the directive. Nevertheless, a more binding formulation of the objective would have been possible, such as the obligation of Europe-wide cross-border compatibility and interoperability of the smart meter framework.

On this basis, the member states have subsequently failed at the intergovernmental level to ensure compatibility and interoperability of smart meters through coordinated implementation of the directive between neighboring states, in this case Germany and France, for example. A characteristic feature here is a consideration of national economic interests and standards that outweighs cross-border interests. This is, of course, typical and can also be observed in other policy areas (e.g., environmental badge).

The smart meter rollout is thus a good example of a missed opportunity at two levels for further technical and legal integration of national legal systems into an overall European system.

References

Assenblée nationale/sénat: Rapport d'information sur les enjeux des compteurs communicants, p 57. Available online: https://www.senat.fr/rap/r17-306/r17-3061.pdf. Jan 26, 2022

Bretthauer S (2017) Smart Meter im Spannungsfeld zwischen Europäischer Datenschutzgrundverordnung und Messstellenbetriebsgesetz, EnWZ 2017, 56 et seq

Bundesamt für Sicherheit in der Informationstechnik: Allgemeinverfügung zur Feststellung der technischen Möglichkeit zum Einbau intelligenter Messsysteme. Available online: https://www.bsi.bund.de/SharedDocs/Downloads/DE/BSI/SmartMeter/Marktanalysen/Allgemeinverfuegung_Feststellung_Einbau_01_2020.pdf;jsessionid=D47E9A0132757BF13B1FC09A51D68239.internet461?__blob=publicationFile&v=1. Jan 22, 2022

Bundesamt für Sicherheit in der Informationstechnik: BSI zertifiziert weitere Smart-Meter-Gateways nach TR-03109-1. Available online: https://www.bsi.bund.de/DE/Service-Navi/Presse/Pressemitteilungen/Presse2022/220208_SMGW_EMH-Theben.html. Feb 10, 2022

Bundesamt für Sicherheit in der Informationstechnik: die Schutzprofile BSI-CC-PP-0073, BSI-CC-PP-0077, BSI-CC-PP-0095. Available online: https://www.bsi.bund.de/DE/Themen/Unternehmen-und-Organisationen/Standards-und-Zertifizierung/Smart-metering/Uebersicht-Schutzprofile-und-TR/uebersicht-schutzprofile-und-tr.html. Dec 9, 2022

Bundesamt für Sicherheit in der Informationstechnik: die technischen Richtlinien BSI-TR-03109, BSI-TR-03109-1, BSI-TR-03109-2, BSI-TR-03109-3, BSI-TR-03109-4, BSI-TR-03109-5, BSI-TR-03109-6. Available online: https://www.bsi.bund.de/DE/Themen/Unternehmen-und-Organisationen/Standards-und-Zertifizierung/Smart-metering/Uebersicht-Schutzprofile-und-TR/uebersicht-schutzprofile-und-tr.html. Dec 9, 2022

Bundesnetzagentur/Bundeskartellamt: Monitoringbericht 2019, recital 10, pp 324–325. Available online: https://www.bundesnetzagentur.de/SharedDocs/Pressemitteilungen/DE/2019/20191127_Monitoringbericht.html. Feb 7, 2022

Chanvry E (2022) La protection des données personnelles avec le compteur Linky. Available online: https://www.hellowatt.fr/suivi-consommation-energie/compteur-linky/donnees-personnelles-protection. Jan 26, 2022

Classen D (2021) Oppermann/Classen/Nettesheim, Europarecht, 9th edn., § 22 recital 18 et seq

Code de l'énergie Art. R341-8 par. 3

Commission Nationale de l'Informatique et des Libertés : Délibération n° 2012-404 du 15 novembre 2012 portant recommandation relative aux traitements des données de consommation détaillées collectées par les compteurs communicants

Commission Nationale de l'Informatique et des Libertés: Linky, Gazpar: quelles données sont collectées et transmises par les compteurs communicants ? Available online: https://www.cnil.fr/fr/linky-gazpar-quelles-donnees-sont-collectees-et-transmises-par-les-compteurs-communicants. Jan 24, 2022

Commission Nationale de l'Informatique et des Libertés: Pack de conformité - les compteurs communicants. Available online: https://www.cnil.fr/sites/default/files/typo/document/Pack_de_Conformite_COMPTEURS_COMMUNICANTS.pdf. Jan 26, 2022

Conrad I (2019) In: Auer-Reinsdorff/Conrad, IT and data protection law, 3rd edn., § 34, para. 16

Deutscher Bundestag: Bundestag-Drucksache 18/7555, p 80, p 108

ECJ 6/64, Slg. 1964, 1251, 1274 seq.—Costa/ENEL

Électricité de France: Fournisseurs d'énergie: les ELD (entreprises locales de distribution) en France. Available online: https://particulier.edf.fr/fr/accueil/guide-energie/demenagement/fournisseur-energie-eld.html. Jan 19, 2022

Enedis: dossier de presse septembre 2021, p 3 and 6. Available online: https://www.enedis.fr/sites/default/files/documents/pdf/DP_Le_compteur_change_pas_notre_engagement_de_service_public.pdf. Dec 15, 2021

Energie le Lynx: Refuser le compteur Linky: le guide ultime. Available online: https://www.lelynx.fr/energie/comparateur-electricite/compteur-electrique/compteur-linky/refuser/. Dec 15, 2021

Energie-Handels-Gesellschaft (2022) Smart Meter—Funktionsweise, Bedeutung, Sicherheit & Rollout. Available online: https://www.eha.net/blog/details/smart-meter-unternehmen.html. Dec 13, 2022

European Commission. Commission recommendation of 9 March 2012 on preparation for the roll-out of smart metering system

European Commission (2022) Smart grids task force. Available online: https://energy.ec.europa.eu/topics/markets-and-consumers/smart-grids-and-meters_en. Dec 8, 2022

European Parliament/Council: Directive 2009/73/EC, recital 27, recital 55, annex I para. 2, Art. 3 para. 11

European Parliament/Council: Directive 2012/27/EU of the European Parliament and of the Council of October 2012 on energy efficiency, amending Directives 2009/125/EC and 2010/30/EU and repealing Directives 2004/8/EC and 2006/32/EC, L 315/1

Gesetz zur Digitalisierung der Energiewende, 29.08.2016, BGBl I 2016, p 2043

Gusy, Eichenhofer. In: Wolff, Brink (eds) Beck OK Datenschutzrecht, 38th edn. as of 1.11.2021; para. 33, para. 78

Haufe: Smart-Meter-Rollout: Habeck plant politischen Neustart. Available online: https://www.haufe.de/immobilien/wirtschaft-politik/smart-meter-rueckt-der-rollout-in-weite-ferne_84342_540252.html. Nov 30, 2022

Hellmuth N, Jakobs E-M. Informiertheit und Datenschutz beim Smart Metering, Zeitschrift für energiewirtschaft 44, p 17

Institut national de la consommation: compteur linky et données personnelles: étude juridique. Available online: https://www.inc-conso.fr/content/compteur-linky-et-donnees-personnelles#:~:text=Par%20ailleurs%2C%20les%20donn%C3%A9es%20personnelles,peuvent%20%C3%AAtre%20identifi%C3%A9es%20qu'indirectement.&text=La%20seule%20information%20qui%20permet%20une%20identification%20est%20le%20num%C3%A9ro%20du%20compteur. Jan 25, 2022

Jahn PB. Stopp für Einbauverpflichtung intelligenter Messsysteme, 109

Jandt S, Roßnagel A, Volland B. Datenschutz für Smart Meter - Spezifische Neuregelungen im EnWG, ZD 2011, 102

Kahl W (2009) Die Kompetenzen der EU in der Energiepolitik nach Lissabon, EUR 2009, 601, 608

Lammel S (2022) Heizkostenverordnung, 5th edn., § 5 recital 70

Lindermann D (2022) In: Säcker, Zwanziger, Berliner Kommentar zum Energierecht, 5th edn., § 49 MsbG recital 10, § 50 MsbG recital 1, § 60 MsbG recital 1 seq

Messstellenbetriebsgesetz vom 29. Aug 2016, BGBl. I, p 2034

Ministère de la transition écologique et solidaire: stratégie française pour l'énergie et le climat: programmation pluriannuelle de l'énergie 2019-2023 2024-2028, p 185. Available online: https://www.ecologie.gouv.fr/sites/default/files/20200422%20Programmation%20pluriannuelle%20de%20l%27e%CC%81nergie.pdf. Dec 15, 2021

OVG Münster: decision dated 4.3.2021 - 21 B 1162/20

Regulation (EU) 2016/679 of the European Parliament and of the Council of 27 April 2016 on the protection of individuals with regard to the processing of personal data, on the free movement of such data and repealing Directive 95/46/EC

Renner S (2011) Smart Metering und Datenschutz in Österreich, DuD 8 2011, p 528

Ruffert M. In: Calliess C, Ruffert M. AEUV Art. 288, recital 97

Säcker FJ, Zwanziger X. Berliner Kommentar zum Energierecht, vol 4. MsbG – Messstellenbetriebsgesetz, 4th edn. Einleitung recital 10, § 30 MsbG, recital 65 et seq

Schweizerische Eidgenossenschaft/Bundesamt für Energie: Energiestrategie 2050 nach dem Inkrafttreten des neuen Energiegesetzes. Available online: https://www.bfe.admin.ch/bfe/de/home/politik/energiestrategie-2050.exturl.html/aHR0cHM6Ly9wdWJkYi5iZmUuYWRtaW4uY2gvZGUvcHVibGljYX/Rpb24vZG93bmxvYWQvODk5Mw==.html. Jan 29, 2022

Sénat: Collecte de données de consommation par des compteurs communicants, 15e législature, Question écrite n° 14568 de M. Cyril Pellevat (Haute-Savoie - Les Républicains) publiée dans le JO Sénat du 05/03/2020, p 1116. Available online: https://www.senat.fr/questions/base/2020/qSEQ200314568.html. Feb 10, 2022

Streinz R (2018) EUV/AEUV, 3rd edn., Art. 28 AEUV, recitals 5 et seq. and 14 et seq

Tribunal de Grande Instance Bordeaux interim order v. 23/04/2019, RG 19/73; 19/75; 19/76; 19/77

Tribunal de Grande Instance Toulouse interim order v. 12/03/2019, RG 19/00431

vom Wege J-H, Wagner F, Ruff H (2017) Die Interimslösung der BNetzA – Anpassung der Marktkommunikation an das Messstellenbetriebsgesetz, IR 2017, 26 et seq

Wieser M (2014) Intelligente Elektrizitätsversorgungsnetze – Ausgewählte Rechtsfragen unter besonderer Berücksichtigung des EnWG 2011 und des EEG 2012, p 48

Wimmer M (2020) Smart Meter, Plattform und Blockchain Datenschutzrechtliche Herausforderungen neuer Digitalisierungskonzepte der Energiewende, EnWZ 2020, 387, 388

Wolff HA, Brink S. Beck'scher Online-Kommentar Datenschutzrecht, 38th edn., status as of Nov 1st, 2019, Einleitung zur DSGVO, recital 20

Technoeconomic Review of Smart Metering Applications

Nikolaos Efkarpidis, Martin Geidl, Holger Wache, Marco Peter, and Marc Adam

This chapter represents a brief version of the survey conducted in Efkarpidis et al. (2022), where various smart metering applications are presented from the point of different stakeholders' interests. In particular, the applications are clustered with regards to three key groups of stakeholders: (a) end-customers, (b) energy service providers, and (c) authorities, and research institutions. The implementation potential for each application is assessed considering the interests and benefits for the key stakeholders, technical and regulatory requirements, as well as limitations and barriers. A business case based on the canvas representation is analyzed for one application, as an example. All business case canvas representations can be found in Efkarpidis et al. (2022). Moreover, the study focuses on the investigation of current business models for smart metering applications. A survey is conducted based on a questionnaire filled by various Swiss stakeholders.

1 Introduction

The metering of energy consumption in electric power systems is conventionally carried out through electromechanical meters that require personnel presence at the meter location for meter readings and other management tasks. Automated meter reading (AMR) was being added to electronic meters between 1970 and 2000, however, only one-way communication was feasible (Efkarpidis et al. 2022).

N. Efkarpidis (✉) · M. Geidl
School of Engineering, Institute of Electric Power Systems, University of Applied Sciences and Arts Northwestern Switzerland, Windisch, Switzerland
e-mail: nikolaos.efkarpidis@fhnw.ch

H. Wache · M. Peter · M. Adam
School of Business, Institute for Information Systems, University of Applied Sciences and Arts Northwestern Switzerland, Olten, Switzerland

D. Ould Abdeslam (ed.), *Smart Meters*, Lecture Notes in Energy 97,
https://doi.org/10.1007/978-3-031-27556-2_7

The energy supplier receives meter reads typically once per month, so there is no need for manual meter reads, as occurred for the electromechanical meters. This ensures timely and accurate billing and allows the customers to analyze their energy usage data. Furthermore, AMR leads to improved security and tamper detection for equipment compared to the conventional meters.

The limitation of AMR systems with respect to one-way communication was overcome by the introduction of smart meters (SMs), which can measure all the electrical parameters like electronic meters and communicate in a bidirectional way (Koponen et al. 2008). SMs are electronic meters used by utilities to exchange information between the energy supplier and the business customer for billing customers and operating the electric power systems. SMs, whether in a power substation or residential setting, enable the nearly real-time measurements needed to monitor equipment health, grid congestion and stability, as well as system operation and control (EU 2019). Moreover, SMs come with an optional smart energy display, which provides information about the real-time energy usage and its cost. While AMR meters can provide the monthly energy consumption and possibly the peak power demand per month, SMs can send considerably more information, including cumulative kWh usage, daily usage, peak power demand, as well as voltage information, outage information, time of use kWh and peak kW readings. They can also send tamper notifications to the energy supplier, and allow physical and wireless connection, as well as bidirectional metering, data storage and management (Obenchain et al. 2011). These meters were first applied to commercial and industrial customers due to the need for more sophisticated rates and more granular billing data requirements, however, they have become available to all consumer classes due to the decreasing cost of technology and advanced billing requirements.

Various definitions have been proposed by national organizations and authorities for the terms of the "smart meter", and the "smart metering systems" or "smart metering infrastructure" (SMI). In Switzerland, international standards and recommendations from recognized specialist organizations were taken into account, when introducing smart metering systems. In accordance with the Electricity Supply Act (Der Schweizerische and Bundesrat 2013), smart metering systems measure the exchange of electrical energy and have bidirectional data transmission to the grid operator. The Swiss law also emphasizes that the term smart metering system includes not only the actual meter itself, but also associated components, such as communication systems, meter data management systems (MDMS) and visualization platforms. The network operator is the operator of the smart metering system. In Germany, the Article 21d of Law on electricity and gas supply defines that a smart metering system comprises the SMs, a central communication unit, named as "smart meter gateway," and a security module (EnWG 2005). Smart metering systems should fulfil, not only the custody transfer requirements, but also the protection requirements and technical guidelines, as defined by the Federal Office for Information Security (EnWG 2005). In summary, a smart metering system consists of three key components:

- SMs installed at the customer's premise to collect the energy consumption data,
- communication network to transmit the data from the meter to the utility office,
- and a MDMS to store and process the interval load data for various control and operation purposes.

Apart from the main components, smart metering systems can also include control technologies, such as programmable communicating thermostats and load control devices for the control of appliances and equipment at the customer's premise. In this context, home area networks (HAN) and energy management systems (EMS) can be included in smart metering systems when automatic control of household appliances is conducted with respect to price signals and load conditions. Information technologies, e.g., web portals, mobile devices, and in-home displays (IHDs) are also used for the visualization of near real-time data about the electricity consumption and costs so that customers will be motivated to better manage their electricity consumption. Figure 1 presents a typical smart metering system, as defined by the U.S. Department of Energy.

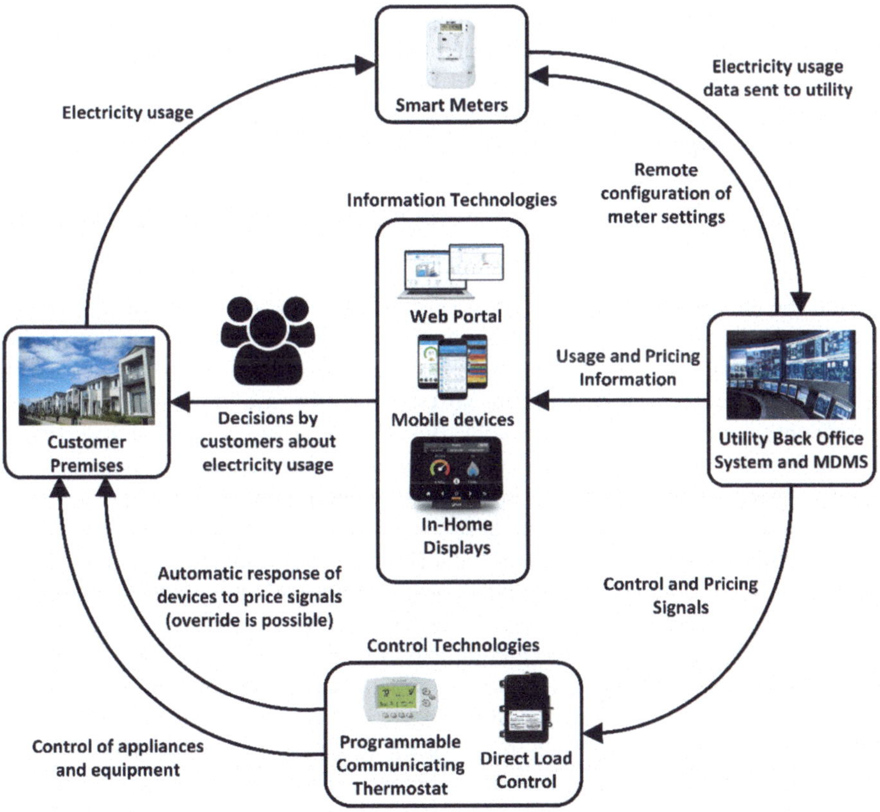

Fig. 1 Typical smart metering system for smart metering applications (U.S. DOE 2016)

It is clear that there is a significant difference between smart meter devices and smart metering systems or SMI. While the smart meter is the individual appliance installed at end-customer's premise, SMI can comprise a large volume of assets, particularly SMs, communication infrastructure, control technologies, and information or visualization technologies. Main prerequisite for the operation of smart metering systems is the presence of an appropriate communication system for data exchange with the utility back office system of the energy provider and the MDMS through the smart meter gateway. A security module is also required for data protection against tampers from non-authorized stakeholders. Finally, visualization platforms, either on the smart meter device or on remote displays are required for monitoring of smart metering data from the end-customers.

2 Smart Meter Roll-Out

According to the Annex I of the Third Energy Package (2009/72/EC) (EU 2009), the European Union (EU) countries shall ensure that the end-customers are actively involved in the electricity market through intelligent metering systems. In particular, as of 2020, 80% of the consumers shall be equipped with SMs, in case that the roll-out in their country is "positively assessed," which means that the cost benefit analysis (CBA) is positive. On occasion of a negative CBA, a reassessment must be carried out every four years. Once the outcome of the CBA is positive, at least 80% of end users must be equipped with SMs within seven years from the date of positive assessment.

In addition to the CBA, other economic, regulatory, and technical barriers have delayed the adoption of smart metering systems. In particular, a serious economic barrier is the sharing of the costs of deploying smart metering systems between the different actors in the industry. The main technical obstacle is related to the current lack of international standardization, so commercially available smart metering components are often not interoperable. Besides that, consumer resistance driven by data security and privacy issues and uncertainty about consumer benefits have negative impact on the adoption of smart metering systems.

As shown in Fig. 2, the smart meter roll-out differs in European countries. More than half of the EU countries have reached a 10% installation rate for electricity SMs. Ten countries have already reached the deployment rate of 80%, while the other EU members will achieve this target by 2025 or later. In Switzerland, a gradual roll-out of 5 million SMs is expected to be completed by 2027, while in France a roll-out of up to 35 million SMs is going to be completed in 2026 (Mora et al. 2019). In Germany, a gradual roll-out of 45 million SMs is expected by the end of 2032. It shall be highlighted that the predictions for the rollout of SMs presented in Fig. 2 come from manufacturer data, thus, these statistics represent market expectations. More conservative predictions by national regulatory authorities can be found in Tounqouet (2019), where it is expected that the penetration rate of SMs will reach up to 84% and 92% by 2024 and 2030, respectively.

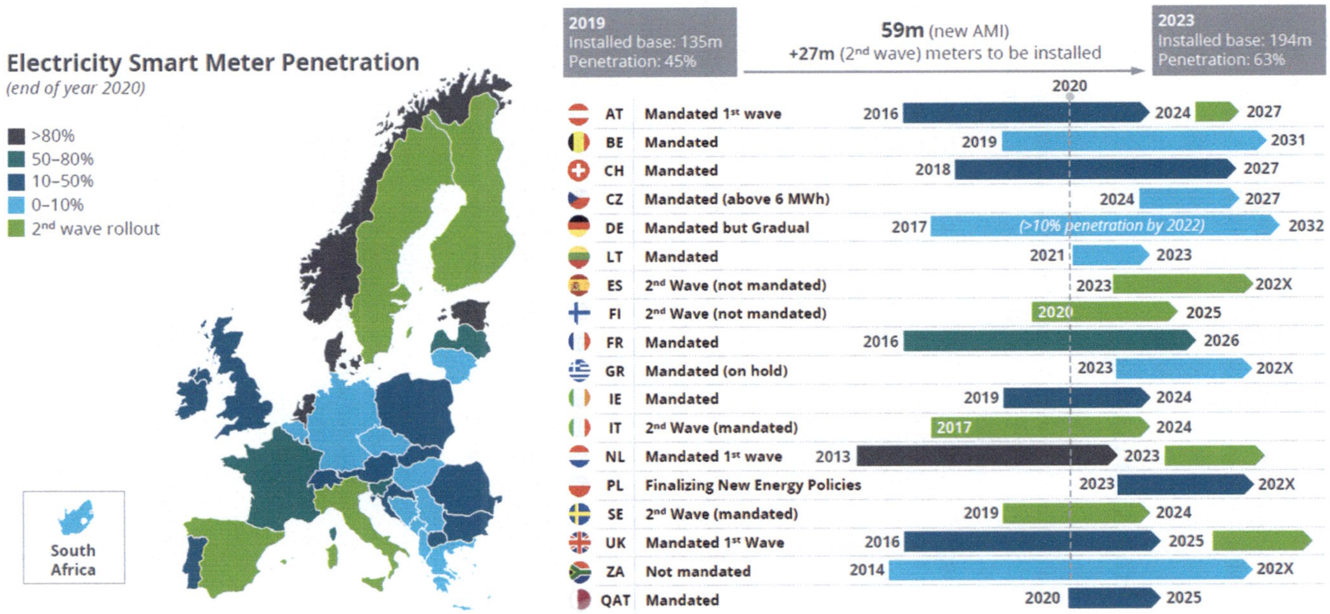

Fig. 2 Roll-out of SMs in Europe by the end of 2020 (Landis+Gyr 2021)

In summary, various reasons, such as negative CBA, unclear legislation to enable smart metering and limited consumer acceptance have delayed the roll-out of SMs in several countries. Nevertheless, the gradual integration of smart metering systems is expected to have an impact on the systematic deployment of various smart metering applications that are still at an early stage of applicability.

3 Smart Metering Applications

Different classifications of stakeholder groups for smart metering applications can be found in the scientific literature based on the actors of bulk generation, transmission, distribution, and energy market levels. In this study, we classify the key stakeholder groups according to the principle of mutually exclusive collectively exhaustive (MECE) so as to limit existing overlaps. The MECE distinction aims to define the stakeholders that are mainly interested in a specific application and make the highest profit. To this direction, the following groups of stakeholders are defined:

- **Customers**: Private, commercial, or industrial end-customer facilities equipped with SMs.
- **Energy service providers**: In this category we combine companies at all levels of the energy supply value chain including producers, network operators, traders, suppliers, balance responsible parties (BRPs), aggregators, etc.
- **Other stakeholder groups**: This class comprises different types of authorities, institutions, and associations that are not directly involved in the business, but they have a substantial interest in smart metering projects. They can include legislative bodies, policy makers, research and educational institutions, public administration, and non-governmental organizations.

The smart metering applications were defined for each of the stakeholder categories, as shown in Fig. 3.

Apart from the technical and functional description of smart meter applications, the associated business models of the applications are also analyzed for selected applications. These models are presented in the form of a business case canvas based on the Business Model Canvas (BMC) from Osterwalder and Pigneur (2010), as shown in Fig. 4.

3.1 Energy Service Provider—Oriented Smart Metering Applications

In this section, the smart metering applications that are of main interest for the energy service providers, are presented.

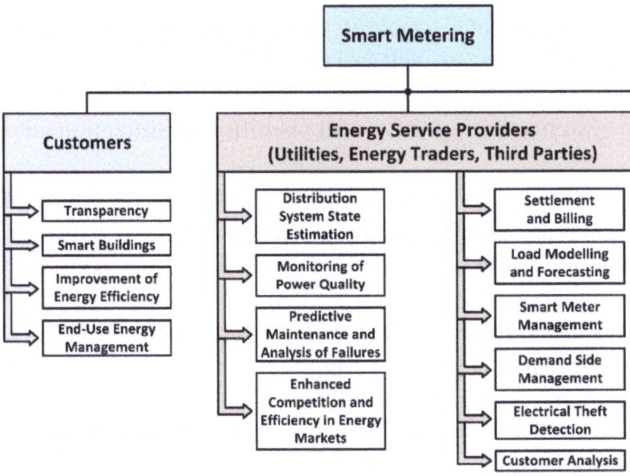

Fig. 3 Existing and upcoming smart metering applications from the point of various stakeholders' interest

Fig. 4 Business case canvas

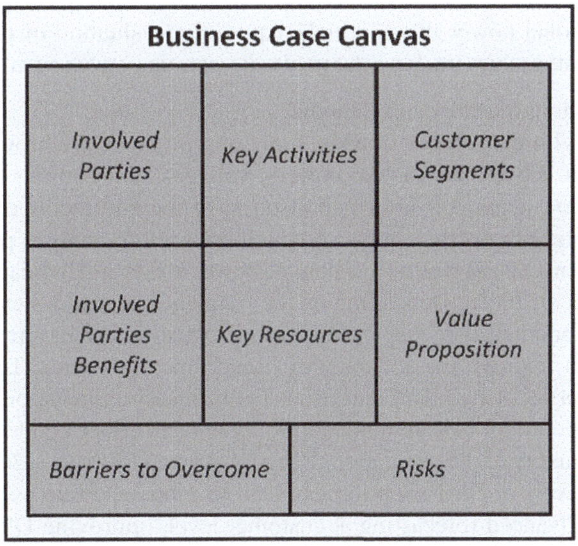

Distribution System State Estimation

State estimation (SE) estimates the steady or dynamic state of power systems based on data provided by measurements and other available system information (Schweppe and Wildes 1970). Static SE provides a snapshot of power system's operating point reflected by state variables, e.g., node voltages and phase angles, while dynamic SE predicts state vectors one timestep ahead based on a priori knowledge and can be corrected according to the subsequent measurement sets. The main objectives of SE are the detection of bad measurement data, topology errors, estimates of network parameters and unmonitored components, as well as smoothing out of small errors (Ahmad et al. 2018). As an along-established tool for bulk grid operation, SE is extensively applied to transmission systems to support system visibility, optimization, and market operation.

The SE in distribution networks, named as distribution system state estimation (DSSE) has not yet seen widespread adoption due to the relatively low degree of monitoring and oversight of distribution systems (Abur and Expósito 2004). In particular, medium voltage (MV) and low voltage (LV) networks are accompanied by less mature communications and telemetry compared to transmission networks, and Supervisory Control and Data Acquisition (SCADA) systems are only applied in MV distribution substations, thus, the state of secondary distribution substations is usually uncertain (Quiles et al. 2012). Due to low redundancy of measurements in distribution networks, pseudo-measurements (e.g., forecasted, or modelled loads according to historical data) are used as supplementary data to increase observability. With the sparse placement of measurement points on distribution systems, the accuracy of DSSE relies heavily on load forecasts to maintain an observable network. Also, distribution networks rely mostly on radial topologies and unidirectional power flows, which require the evaluation of peak loads and fault currents, rather than tracking the network operating state in real-time.

Main Interests and Benefits

With the roll-out of SMs, more opportunities are provided to utilize measurements of active and reactive powers, voltages, and currents at various points of distribution networks. Information on such measurements can be handled by the DSSE algorithms, the results of which are used to analyze permissibility of voltages and currents, as well as to calculate power losses and indicators of grid reliability (Wakeel et al. 2016). Depending on the frequency of smart metering data updates, the corresponding data can be used as near real-time measurements for dynamic DSSE or to improve the accuracy of pseudo-measurements (Liao and Milanović 2016). In terms of bad data detection, faulty measurements, or temporarily bad data due to recording or communication failures can easier be identified, enhancing the overall quality of the estimation (Razanousky and Morrissey 2018). Furthermore, even small levels of SMI adoption can lead to more accurate pseudo-measurements owing to advanced forecasting at customer level, improving DSSE effectiveness. The DSSE can consider all types of available measurements, reducing the investment costs into the minimum required measurement infrastructure, since estimates of the grid state are also provided by DSSE for the nodes where SMs are not located (Liao and Milanović 2016).

With more accurate real-time model of the network through the DSSE, various additional applications of the distribution management system (DMS), e.g., distribution contingency analysis, support asset management, optimal network reconfiguration, fault location and restoration, voltage control, can be more reliably performed and controlled, as shown in Table 1. Hence, DSOs will obtain an improved view of the current grid state. In addition, customer demands at power peaks can be reduced. Improvements in utilization of grid assets and contingency planning can also be achieved, deferring any grid upgrades. Finally, more detailed models can be obtained for grid planning purposes and for the connection of new loads.

Regulatory and Technical Requirements

It is evident that an DSSE tool can generate good quality estimated values when provided with sufficient measurement data. Consequently, adequate instrumentation and communication support are crucial for the reliable operation of the DSSE tool. In other words, the deployment of the required number of SMs on distribution networks is a prerequisite for the satisfactory performance of DSSE (Primadianto and Lu 2017; Kemal et al. 2020). The provision of near real-time data is required for the grid state estimation in real-time operation (Kemal et al. 2020). Besides the frequent provision of smart metering measurements, accurate topology data are also needed for the DSSE (Primadianto and Lu 2017). In particular, incorporation of SMI and other customer-side measurements requires accurate modeling of the associated distribution network. Another important requirement is to ensure that the readings from SMs and other existing systems (e.g., SCADA) are synchronized (Primadianto and Lu 2017).

Barriers and Limitations

While there are significant opportunities for further deployments in this area, the scope for using smart meter data as input to DSSE for real-time applications is limited for a number of reasons including data privacy issues, low smart meter data, low bandwidth of communications technologies, low data reliability and limited data synchronization (Razanousky and Morrissey 2018; Primadianto and Lu 2017; Kemal

Table 1 Features and benefits of advanced DMS applications through the use of DSSE (Schweppe and Wildes 1970)

Application	Functionality	Benefits
Unbalanced load flow analysis	Definition of line currents and node voltages per phase for the distribution grid	• Improved system awareness • Higher asset utilization • Improved contingency planning
Fault location	Identification of possible fault locations on system	• Improved crew efficiency in outage management • Reduced SAIDI[a] and CAIDI[b]
Distribution voltage control	Monitoring/control of grid components to reduce peak load and losses	• Reduce customer demand at power peaks • Lower system losses • Improved voltage profiles
Remote switching and restoration	Feeder reconfiguration considering grid status	• Reduced SAIDI and CAIDI • Lower system losses

[a] System Average Interruption Duration Index

[b] Customer Average Interruption Duration Index

et al. 2020). Today, many DMS operators can only expect to have access once-a-day, or few-times-a-day values of energy consumed by customers equipped with SMs for data privacy and security reasons (Razanousky and Morrissey 2018; Primadianto and Lu 2017). There may be also lack of sufficient telemetry infrastructure to relay data points back to the control center, and many systems do not have the bandwidth to transmit measurements to the data hub without significant delay (Primadianto and Lu 2017; Kemal et al. 2020). In addition, the instrumentation of current distribution grids consists of limited real measurements, accompanied by some virtual- and numerous pseudo-measurements to achieve the required observability (Wakeel et al. 2016; Liao and Milanović 2016). The degree of trust in the available information varies, depending on the type of data and the accuracy of instruments, and is highly diverse. Inaccurate data of the distribution grid e.g., line impedances, load phase connection or grid topology, as well as bad and missing data can additionally contribute to the inaccuracy of DSSE (Primadianto and Lu 2017). Time synchronization in taking the measurement samples at different smart meter locations is also very crucial, since non-synchronous measurement data can increase the error margin of the DSSE output (Primadianto and Lu 2017). The smart meter measurements are usually asynchronous with the more real-time SCADA measurements, since the built-in real-time clock of SMs is only periodically synchronized with actual time (Weranga et al. 2013; Razanousky and Morrissey 2018). As a result, DSSE may be executed using a set of measurements many of which are delayed by anywhere from a few seconds to days (Alimardani et al. 2015; Peppanen et al. 2015).

Relevant Instances
Several research projects have been recently aimed at developing DSSE techniques based on the use of SMs, as shown in Table 2. Authors in Pokhrel et al. (2019) assess the performance of a DSSE approach on a MV grid based on real-time measurements from SCADA systems embedded with SMs and conclude that the selection of the most crucial smart metering measurements has direct impact on the high network observability, minimum measurement utilization and grid state estimation. In Waeresch et al. (2015), the DSSE architecture allows the exchange of data among the different voltage levels and permits the interaction with other services that use the results of DSSE as input for their processes. A crucial procedure for the high accuracy of DSSE is the bad data analysis that can be considerably improved with the increased smart metering measurement redundancy in the grid providing that the accuracy of pseudo-measurements is also sufficient (Waeresch et al. 2015). Nevertheless, in Ulbig et al. (2016), smart metering redundancy has limited impact on the DSSE performance, since measurements from two out of 13 buses are required for voltage errors below 0.5% in the investigated Kleinhüningeranlage LV grid of IWB. Apart from the smart metering redundancy, the accuracy of state estimations depends on the completeness of all loads in the grid, as well as the topology and the cable parameters assumed for the examined network (Herbst et al. 2019). In the context of Upgrid project, in Lisbon, three months of historical measurement data from 54 SMs

installed at the LV customers' side were utilized for grid state estimation (Moreira et al. 2020). For the investigated scenarios, the maximum absolute error was equal to 0.66 V for the examined months, which makes it suitable for real-time simulations. Finally, during the preparation of information for state estimation of LV distribution grids, the main barriers can be related to the identification of phases of consumers' connection, synchronization of smart metering measurements, and errors in information about the grid topology and the connection of LV feeders to transformers (Kuzkina and Golub 2018).

Business Case Canvas

Distribution system state estimation proposes several benefits to the DSO (customer segments), such as the detection of critical parts of the power grid. Involved parties are only the end-customers (as "data provider"), as they can detect problems or errors in the grid much more easily. But the main requirement is also the acquisition of real-time measurements with additional devices. The business case canvas for this application with its details is shown in Table 3.

Table 2 DSSE projects based on SMI (Pokhrel et al. 2019; Waeresch et al. 2015; Moreira et al. 2020; Pau 2017; Kuzkina and Golub 2018; Herbst et al. 2019; Ulbig et al. 2016)

Project name	Pilot description
Denmark DECODE	Demonstration on a model of a 52-bus MV distribution network from the Lind area using SMs, PMU[a], RMU[b] etc. (Pokhrel et al. 2019)
Germany SmartSCADA	Evaluation of the LV state estimation algorithm based on SM data and PV-feed-in forecasts for a typical LV grid from the DSO Stadtwerke Kaiserslautern with 120 loads and 24 PV systems (Waeresch et al. 2015)
Portugal Upgrid	Historical smart metering data from 54 SMs installed at the LV customers' side were used for grid state estimation considering both three-phase and single-phase end-users (Moreira et al. 2020)
Sweden FLEXMETER	Assessment on a sample distribution grid, composed of a 15 kV network and 400 V networks by using measurements from the SMs at the final customers' premises (Pau 2020)
Russia SB RAS	Demonstration on a typical Russian LV grid of the Iskra village consisting of 157 nodes, where the sufficient number of measurements from SMs or AMI is examined (Kuzkina and Golub 2018)
Switzerland SCCER	Estimation of voltage drops in a LV distribution network of Arbon city based on measurements from SMs installed at end-customers' premises (Herbst et al. 2019)
Switzerland SFOE Project	Evaluation of the DSSE algorithm on a Kleinhüningeranlage pilot grid using smart metering measurements provided by the utility of Basel, Industrielle Werke Basel (IWB) (Ulbig et al. 2016)

[a] Phasor Measurement Unit

[b] Remote Measurement Unit

Table 3 Business case canvas: distribution system state estimation

Involved parties • End-customers installing a SM • DSOs	*Key activities* • Analysing smart meter data • Support system visibility, optimization, and market operation • Near real-time measurements	*Customer segments* • End customers
Involved parties benefits • Reliable power supply • High availability of the grid	*Key resources* • SM data • DSSE algorithms	*Value proposition* • Improved distribution system state estimation • Better control, and management of the distribution networks • Smoothing out of small errors • More opportunities are provided to utilize measurements of active and reactive powers, voltages, and currents at various points of distribution networks • Bad data detection, faulty measurements, or temporarily bad data due to recording or communication failures can easier be identified, enhancing the overall quality of the estimation • Improve the accuracy of pseudo-measurements/forecasts • Unbalanced load flow analysis • Fault location • Distribution voltage control • Remote switching and restoration
Barriers to overcome • Real-time measurement • Many measuring devices are needed • Existence of bad data to be removed • Identification of phases of consumers' connection • Synchronization of smart metering measurements • Errors in grid topology and the connection of LV feeders to transformers • DSSE has not yet seen widespread adoption due to the relatively low of monitoring and oversight of distribution systems		*Risks* • None to the business case

Settlement and Billing

Settlement is the process by which the actual electricity consumed by the end-customer is determined in order to be used for billing procedures (CMA 2015). Billing procedure based on settlement is periodically carried out and shows the balance amount based on the difference between the actual amount of electricity consumed throughout the year and the total amount paid during the same period (EC 2018). In some countries, there are also meters with the option to pay in advance, the so-called "prepayment meters," for purchasing electricity. While older installations are equipped with electromechanical electricity meters and coin-operated machines, more modern electronic meters are implemented in combination with chip card readers.

With the use of SMs in settlement process, consumers can receive more accurate and timely bills (Landis+Gyr 2014; ERGEG 2007). Settlement procedures will be carried out remotely eliminating field visits for meter readings (Landis+Gyr 2014; Owen and Ward 2008). Smart metering data can also be used for more frequent billing, helping end-users manage better their consumption (Landis+Gyr 2014; ERGEG 2007). SMs can achieve more streamlined switching procedures, as customers will no longer have to deal with adjusted billings, when they move in or out of premises (Landis+Gyr 2014; Morch et al. 2007). In case of prepayment metering, SMs perform more cost-efficient, flexible, and customer-friendly than the conventional prepayment meters (Souvnik 2014). In particular, customers can receive alarms when the available credit reaches a threshold reducing disconnection issues. The prepayment accounts can be recharged by using SMS, phones, web or call centers owing to the two-way communication of SMs with other smart devices (Owen and Ward 2008; Souvnik 2014).

The main technical prerequisite for the participation of SMs in settlement process is the short settlement time, hence, the frequent provision of SM data to the energy supplier and/or the third party that takes over the settlement and billing procedures (Tounqouet 2019; CMA 2015). Frequent billing based on SM data requires the use of data storages for meter readings and reliable communication network (World Bank Group 2018a; Price et al. 2012). When switching a supplier, requirements of interoperability should be fulfilled for successful access and data exchange between the SM and the new supplier (Tounqouet 2019). As for prepayment smart meters, suitable platforms are required for monitoring the balance amount, e.g., using mobile phones, and in-home displays (IHDs), calling the support center and checking the meter (Souvnik 2014).

With regard to the prompt reading of SM data with a high temporal resolution, there are restrictions in many countries for reasons of data security. For the implementation of an automated end-to-end process, the companies involved have to put up with the corresponding expenses for hardware and software (World Bank Group 2018a; Price et al. 2012; ELEXON 2010). Technical restrictions exist in the area of communication technology with regard to bandwidth and availability (World Bank Group 2018a). Interoperability can be a challenge when changing suppliers of components, as well as involving third parties (e.g., aggregators) (World Bank Group 2018a;

ELEXON 2010). One of the main challenges also regards the high level of IT security for the entire smart metering system. On the one hand, the security issue raised during billing based on smart metering is related to the manipulation of measurement data (Zabkowski and Gajowniczek 2013). On the other hand, privacy concerns exist since the electricity consumption profiles may reveal sensitive information about the end-customers (Jawurek et al. 2011).

Predictive Maintenance and Analysis of Failures

Preventive or time-based maintenance is applied to ensure that an asset is inspected and serviced before it reaches the point of failure, hence, it is typically scheduled based on operating experience with the asset class (Khuntia et al. 2016; Reed 2018). Predictive or condition-based maintenance requires continuous monitoring of asset performance through sensor data and prediction engines to provide indication of equipment issues and failures (Reed 2018). Therefore, predictive maintenance typically uses advanced pattern recognition and predictive data analytics for real-time insights of asset health. Predictive maintenance enables each asset to be served only when required, hence, asset service can be carried out before the regular preventive maintenance. SMI measurements, e.g., consumption and event data, can be used for predictive maintenance and analysis of failures (Moghaddass and Wang 2018).

With the usage of SM data, utilities can monitor the conditions of various assets (e.g., transformers) in near real-time, developing new asset management strategies and optimizing asset lifecycles (Tripathy 2017). Abnormal conditions can proactively be identified, and hence, actions can be appropriately prioritized to prevent power delivery disruptions and optimize overall grid reliability (IBM 2012). Life expectancy of assets can considerably be improved, reducing the total costs, and achieving long-term planning (World Bank Group 2018a). In case of failure analysis, SMI can lead to earlier event detection, faster reporting, easier alarm management and shorter recovery or restoration time (Atos Worldgrid 2010). SM data combined with geographic information system (GIS) information can reduce the average power outage duration due to their fast response and rectification to power outages and faults (Tram 2008).

One of the main prerequisites for the high performance of advanced data analytics applied for predictive maintenance is the integration of large and diverse datasets that are used not only as a training basis, but also as verification data for the developed algorithms (Zhang et al. 2018; Kinetica 2018). Apart from smart metering measurements, additional data from electricity network, meteorological information system and GIS are required, e.g., feeder composition, asset maintenance history, data from grid-edge sensors, as well as weather forecast data, calendar information and historical outage details (World Bank Group 2018a; Tripathy 2017; Zhang et al. 2018). In addition, reliable communication networks, appropriate data storage infrastructure and information management programs, as well as advanced grid-specific analytics are required for the predictive maintenance and the analysis of failures (Moghaddass and Wang 2018; Hejazi and Rad 2018; Vié et al. 2015).

Primary reasons for the limited applicability of predictive maintenance and analysis of failures with smart metering data include the high costs and the large volumes of high-velocity data collected from various sources (Moghaddass and Wang 2018; IBM 2012; Zhang et al. 2018; Kinetica 2018; Vié et al. 2015). In particular, high investments are required in data storage infrastructure, communication networks and information management programs (Moghaddass and Wang 2018; Hejazi and Rad 2018; Vié et al. 2015). Apart from complexity, data access and privacy issues are also cited at the top implementation challenges, as data cannot be pooled by data storages for the benefit of the utility, while data privacy concerns security and confidentiality of sensitive data (Vié et al. 2015). Finally, the applied methods demand long training time due to their multi-parametric nature, while inaccurate or insufficient data can be a bottleneck on predictive maintenance or analysis of failures (World Bank Group 2018a; Moghaddass and Wang 2018).

Monitoring of Power Quality

Power utilities are subject to technical requirements for the quality of their services including continuity of electricity supply and power quality laid out in national law, standards, and grid codes. Most electrical equipment used by the customers are designed to operate within a narrow band of voltage and frequency, thus, any deviation from the rated band may lead to deterioration of the equipment performance.

Power quality analyzers are typically used to monitor the voltage quality at neuralgic points of the distribution networks, such as industries, renewable energy sources (RES)-based plants or other large-scale power sites. SMs enable continuous monitoring of power quality in multiple network points, preventing any violations of voltage statutory limits, as well as any damages to the grid or the customers' side (Albu et al. 2017). Based on SM data, availability indices, e.g., SAIDI and CAIDI, can be determined automatically and in near real-time, providing early warning or alarms in case of power supply interruptions. SMs can also be used to detect and record current transients and harmonics, voltage sags and swells, flicker, asymmetry, or other power quality issues (ENA 2012). Moreover, SMs can help utilities optimize voltage levels on distribution feeders and reduce grid losses. Near real-time power quality data can also be utilized as performance-based rates for the control of loads. As indicators of possible power quality disturbances in the network, SMs can also help in classifying the disturbances and localizing their sources (Music et al. 2013; Borges et al. 2016; Parvez et al. 2019; Chang et al. 2017).

SMs provide reliable information on voltage quality, providing that they are compliant with specific standards, such as the EN 50160 (CENELEC 2010) and the IEC 61000-4 (CEI/IEC 2003). The main technical requirements associated with the smart metering systems that are also used for power quality monitoring regard: (a) transfer and storage of data, (b) modular and flexible functionality for SMs, and (c) intelligent data management and distributed data storage (Maheswaran et al. 2015; Campbell et al. 2015). The efficient transfer of data requires effective communication links, smart systems able to communicate with devices in order to download information and verify their availability, as well as robust information systems for gathering and analyzing data with minimal human intervention (ERGEG 2009). In addition, interoperability of equipment from different vendors should also be ensured (Zhang et al. 2018). Time sequencing is also essential for correlating power quality events between different instruments, as the start and end time of the measurement intervals needs to be synchronized.

The main barriers for using SMs for power quality monitoring purposes are related to capabilities of SMs, limitations in data storage capacity, bandwidth throughput of the SMI communication infrastructure, and lack of interoperability (Albu et al. 2017; Music et al. 2013; Maheswaran et al. 2015; Campbell et al. 2015). Power quality monitoring of all grid nodes is from an economic point of view not feasible, and not necessary from a technical point of view. Thus, this functionality can only be used to a limited extent at the key grid nodes, e.g., nodes of connection to the transmission grid, MV/MV and MV/LV transformer stations, as well as customers sensitive to power quality disturbances and consumers that significantly affect power quality (Ali et al. 2016).

Load Modelling and Forecasting

Load modeling is essential to power system analysis, planning and control, thus, various load models, e.g., constant power (P), current (I), impedance (Z) or their combination (ZIP), are widely used depending on the application (Khan et al. 2018). However, the integration of new technologies, e.g., heat pumps and distributed energy resources (DERs), have increased the uncertainty and difficulty in load modeling and forecasting (Khan et al. 2018; EPRI 2011; Arif et al. 2018). With the help of SM data, the basis for new deterministic and stochastic models as well as for machine learning (ML) applications can be established. In terms of load modelling, more accurate models can be developed with SM data since the load composition estimation for component-based load modelling can be improved (EPRI 2011). Conventional load models and forecasts use aggregated measurement data or selective measurements, e.g., on transformer level. The high granularity of SM data offers great potential for improving aggregate forecast accuracy (Wang et al. 2019). Utilities can also monitor how customer patterns in device and energy usage evolve over time (Jonhson et al. 2018). The combination of SMI with equipped sensors can also enable the definition of meaningful interdependencies between the load variation and various impact factors, e.g., weather situation, calendar, or geographic data (Khan et al. 2018). At end-customer level, SM data can also be used for the validation of residential load profiles generated by bottom-up approaches (Chuan and Ukil 2015). Furthermore, fine-grained SM data let utilities estimate better system peaks and disaggregate system loads into meaningful customer and regional peaks (Jonhson et al. 2018). The behavior of individual loads in the grid can be modelled and forecasted sufficiently when the influencing variables on the load are known. In this context, the collection of "ground-truth data," such as, calendar factors, weather and energy trading conditions, specific appliances' demand, is required for the development of ML methods. When

ZIP models are used, time-synchronized and high-resolution voltage measurements are also needed. The ML algorithms should be robust to bad data, missing measurements, and noises, thus, data preprocessing steps, must usually be undertaken in order to provide a consistent data feed for the training and predictive models (Noh and Rajagopal 2013; Day et al. 2014; Vazquez et al. 2017). Due to the high granularity of SM data, they shall be anonymized or semi-anonymized for data privacy and protection reasons (Asghar et al. 2017). A different private metering process based on aggregated load profiles may also be applied for load forecasting purposes (Eibl et al. 2018). Since the load reading resolution of current SMs ranges from 15 min to 1 h, the effects from instantaneous load changes and from system voltage deviations to active and reactive powers of the load cannot be distinguished. In addition to smart meter data, end-use data for each load component is also required, however, it is difficult to monitor multiple individual loads in practice (EPRI 2011). The smart meter data provides an opportunity to predict the load at different hierarchical system levels, however, the smaller the system, higher volatility in profiles lead to more uncertain results (Khan and Jayaweera 2019). In addition, it is challenging to find a meaningful relationship between load variation and impact factors. Additional barriers of load forecasting and modeling based on smart metering data are related to the large number of datasets and the considerable computational time and effort (Arif et al. 2018; Chaouch 2014).

Customer Analysis

Customer analysis is related to the load profiling that develops basic electricity demand patterns for each consumer or a group of consumers. Traditionally, end-customers are divided into more or less consistent groups based on the capacity of their grid connection and annual electricity consumption. However, the customers' load patterns belonging to the same type of activity may exhibit considerable differences (Figueiredo et al. 2005). The knowledge of electricity consumption patterns is a key aspect for the energy suppliers, so as to obtain consistent groups of consumers with similar characteristics (Cerquitelli et al. 2018). This classification serves as the basis for the design of offers and tariffs.

Fine-grained SM data can be used for load analysis, as the energy consumption pattern of each customer can be defined with a high time resolution on a daily basis. In addition, consumption data combined with information about the type of consumer are used for the construction of user profiles that can be representative for the type of user (Yildiz et al. 2018; Grigoras et al. 2014). The dependence of the load on various influencing factors, e.g., temperature and solar radiation, can also be modelled with available SM data (Chitsaz et al. 2015; Li et al. 2017). By applying clustering techniques to identify groups of end-customers with similar load patterns from SM data, the accuracy of the system level intra-day load forecasts can also be improved (Khan and Jayaweera 2019; Chaouch 2014). Energy consumption analysis at the individual customer level can also help the energy suppliers to identify which characteristics correlate with energy behavioral use (Haben et al. 2016; McLoughlin et al. 2012). Furthermore, pricing can be optimized by designing more suitable tariff options (Flath et al. 2012), while customers viable for energy saving solutions can be identified (Kwac et al. 2014).

One of the main prerequisites for efficient customer analysis is the balance of training datasets, as there may be some attributes that are under- or over-represented (Carroll et al. 2018). In addition, the length of the time series should be carefully selected, as larger sample sizes yield better population estimates with lower variability. Apart from the collection of fine-grained SM data, a key aspect of customer analysis is the selection of the correct attributes for clustering purposes, e.g., outdoor temperature, socio-demographics, and electricity consumption patterns (Prado et al. 2020; Kwac et al. 2013; Beckel et al. 2015). Due to the large volumes of SM data used for customer analysis, preprocessing steps should be necessarily conducted so as to transform the data into a useful format (Carroll et al. 2018). Apart from the preprocessing steps, a set of appropriate ML techniques can be applied the customers' consumption patterns (Grigoras et al. 2014; Kwac et al. 2014; Carroll et al. 2018). Data protection and privacy aspects shall also be considered when utilizing fine-grained smart meter data and other customer attributes.

Customer analysis based on smart metering data can be challenging due to the major reasons of high dimensionality, incomplete data coverage and errors in data (Wijaya et al. 2014; Kim et al. 2016). Due to the humongous scale of SM data, many algorithms become intractable when the input is high-dimensional, increasing the required amount of training data and the training time (Carroll et al. 2018). Missing, duplicate or noisy data can also be barriers for the analysis of customers' consumption profiles. Clustering the customers' consumption profiles is also challenging, as the electricity consumption depends on various behavioral factors, e.g., occupancy and willingness to use certain appliances (Cerquitelli et al. 2018).

Enhanced Efficiency and Competition in Energy Markets

Thanks to the energy legislations adopted over the years, electricity markets undergo changes shifting away from a monopolistic environment dominated by a few companies to an increasingly competitive environment (Bogdanovic et al. 2020; EC 2016). Greater efficiency through competition should lead to lower costs and hence prices, than would otherwise have been the case. Future energy markets could even enable consumers to participate in certain market actions. This new structure can be facilitated by smart metering systems that enable near real-time and remote control of consumer devices (Zepter et al. 2019; World Energy Council 2018).

By using up-to-date consumption data from SMs, the total consumption can be better predicted reducing the imbalance costs for the electricity suppliers, and the charges in electricity bills for the customers. More cost-effective contracts can also be offered for potential end-customers, while prosumers can be engaged in various potential markets, e.g., peer-to-peer markets, prosumer-to-interconnected or "island" mode microgrids and organized prosumer groups (Parag and Sovacool 2016). SMs can be used as a prerequisite to introduce new tariffs that promote self-consumption, reduce network usage, and provide economic signals that are consistent with energy markets (Tounqouet 2019). End-users that modify their consumption based on SM data can also enhance price elasticity reducing the risk for electricity market failures and collusive market behavior (Thimmapuram and Kim 2013). As SMs can simplify the switching procedures, customers will be motivated to look for financially effective contracts, increasing the competition among the suppliers (Owen and Ward 2008).

The minimum functional requirements of SMs for their participation in energy markets are mainly referred to the metering frequency, settlement period, parties that access the metering data, communication interface and load control abilities (USmartConsumer 2014; EURELECTRIC 2017). Smart metering systems shall enable customers to be metered and settled at the same time resolution as the national imbalance settlement period (Tounqouet 2019; EURELECTRIC 2017). Access and exchange of data through an appropriate information and communication infrastructure (ICT) is needed for the customer and eligible third parties, as data used for settlement purposes shall be certified by an independent third party (Vos et al. 2018). Furthermore, various incentives, e.g., explicit bill credits or payments for pre-contracted or measured load reductions are also necessary for the engagement of end-customers in energy markets (Eissa 2011).

SMs are mostly not able to dynamically identify a particular time or set certain hours as critical, rendering the energy management systems necessary to identify opportunities for energy savings (EURELECTRIC 2017). Lack of transparency in supply pricing can be considered as a major barrier for the energy suppliers that want to offer dynamic pricing schemes to their customers (CEC 2014). Lack of standardization in fundamental parts of hardware and ICT infrastructure in smart metering systems can also cause communication issues and insufficient interoperability. One fundamental obstacle for the usage of SM data in energy markets is also the utilities' risk aversion that generally leads to slower adoption of new technologies (CEC 2014). As for the economic barriers, the apportionment of cost and benefits among different exchange partners remains a challenge (Nursimulu 2015). Finally, the main social barriers to customer engagement are the lack of awareness, and willingness on the part of the customer (CEC 2014).

Demand Side Management

Demand Side Management (DSM) is the planning, implementation and monitoring of utility activities that are designed to influence customer use of electricity with the main goal to produce desired changes in the utility's load (Gellings 1985). The key to make DSM effective is to integrate fully and dynamically consumer's loads, and information about their usage into the operation of the grid. Nowadays, the development of SMs, home automation, as well as advanced communication and control technologies enables more sophisticated DSM forms even at the household level, with domestic customers being able to adapt demand at their discretion in response to time-varying price signals.

Based on SM data, DSM can offer benefits on system operation, system expansion and market efficiency (Shen et al. 2012; Chatterjee et al. 2019; DOE 2006; Conchado and Linares 2012). SMs enable energy suppliers to deploy DSM programs that encourage actions by customers to modify their electrical usage and consequently, to reduce the peak load and network losses, improving the power generation efficiency (Shen et al. 2012; Conchado and Linares 2012; Balmert and Petrov 2010). Lower electricity usage in peak periods can also improve the utilization of generation and transmission assets, while grid congestion situations can be better handled with lower risk for outages (DOE 2006; Conchado and Linares 2012). SMs may allow the DSO to switch loads in emergency situations utilizing a customer's priority list, excluding emergency service providers and critical customers from switching (DOE 2006; Siano 2014). Based on gathered SM data, controllable loads can be selectively addressed allowing DSOs to maintain grid stability and power quality without building expenditure on additional capacity infrastructure (Conchado and Linares 2012; Siano 2014). Customers that respond to price signals can reduce their consumption increasing their savings in electricity bills, while the cost of electricity production also drops holding down prices in electricity spot markets (DOE 2006; Conchado and Linares 2012; Siano 2014).

A plethora of technical and regulatory requirements should be met for the utilization of DSM programs based on smart metering data. Technical requirements are mainly related to the frequency of meter data readings, incentives for consumers, as well as role and responsibilities of possible DSM aggregators and compensation/remuneration for DSM flexibility. Besides that, regulatory requirements based on the data security and protection aspects should also be fulfilled by the stakeholders involved in DSM.

From the technical point of view, there is still uncertainty for the DSM potential when using SM data due to complexity of both power system operations and advanced DSM forms (Nursimulu 2015). In most cases, there is lack of ICT and metering infrastructure required to deploy DSM. In addition, the size of data, high dimensionality and heterogeneity of the load profiles pose great computational challenges to DSM at consumer level (Khan and Jayaweera 2019). In terms of economic barriers, the heavy investment in ICT and metering infrastructure required for the large-scale involvement of consumers demotivates the utilities to invest due to uncertain revenues (Khan and Jayaweera 2019; Nursimulu 2015; DOE 2006). There is often also not a clear business case for small consumers to participate in DSM due to uncertainties about the costs and benefits, including future revenue streams (European Smart Grids Task Force Expert Group 3 2019). Social barriers are mainly related to the lack of awareness and trust, as well as consumer dissatisfaction from continuous load reduction requests or reduced comfort (Nursimulu 2015; European Smart Grids Task Force Expert Group 3 2019). Finally, regulatory issues arise mainly from missing legal aspects, standardization, and data protection (Nursimulu 2015; European Smart Grids Task Force Expert Group 3 2019).

Electrical Theft Detection

Electrical theft is one of the most prominent issues pertaining to conventional power grids and has been a major concern to the utilities for quite a long time. According to Northeast Group LCC (2017), about $96 billion are lost every year worldwide due to electricity theft. The losses caused by electrical theft are generally external to power systems and referred as non-technical losses, which can be directly computed or precisely estimated, as they may be caused by unknown behaviors that are not accounted by DSOs (Messinis et al. 2013). Electrical theft mainly refers to intentional and illegal use of electricity by various means, e.g., physical attacks in electromechanical meters or cyber-attacks in various IoT devices integrated in electric power systems (Otuoze et al. 2019; Czechowski and Kosek 2016; Polgári and Raisz 2012; Mohassel et al. 2014; McLaughlin et al. 2013; Singh et al. 2019; Jokar et al. 2016).

SMs allow the utilities to detect any abnormal consumption by identifying meter manipulation or implausible measurements. On the contrary to electromechanical meters, various physical attacks, such as, direct connection to distribution lines, keeping open the neutral wire, removal of meter cover, can be eliminated by SMs (Otuoze et al. 2019; Czechowski and Kosek 2016). SMs use data loggers that are digitally controlled and capable of recording tamper-suspicious events, such as, voltage drops, outages and power flow inversion (Polgári and Raisz 2012). If obvious tampering logs have been registered, the SM can immediately alarm the center to initiate a remote switch off to prevent the fraud (Polgári and Raisz 2012). In terms of cyber-attacks, different types of techniques, e.g., state-based, game theory-based and artificial intelligence (AI)-based, have also been examined for theft detection using SM data. The main requirements related to the electricity theft detection through SMs regard the equipment, data, and detection methods. As for the smart metering infrastructure, various hardware components shall be considered for tampering, e.g., tamper-proof enclosure, physical locks, alarm, and self-integrity systems (Mohassel et al. 2014; McLaughlin et al. 2013). Concerning the data requirements, data acquisition is necessary from SMs, as well as substations or distribution transformers for the detection of energy balance within the grid (Sahoo et al. 2015; Kadurek et al. 2010; Berg et al. 2011). Hence, meter audit logs of physical and cyber-attacks should be combined with consumption data for more accurately modeling and detection of theft-related behavior. Regarding the theft detection techniques, specific requirements should be considered for each of the aforementioned three categories (McLaughlin et al. 2013; Singh et al. 2019; Jokar et al. 2016). Though attackers targeting meters have to erase all traced of logged events stored in the SMs, "cyber-tampering" of smart metering systems is not completely impossible (McLaughlin et al. 2013). Smart metering data can be tampered inside the meters when data are stored, and over the communication links, by modifying the meter firmware/storage, stealing credentials to login to meters, as well as exhausting CPU/memory, intercepting the meter communications and flooding the WAN bandwidth (Jakaria et al. 2019). When state-based methods are used for theft detection, high investment costs are required for the monitoring system. Issues with AI-based methods regard data imbalance, vulnerability to contamination attacks, plethora of non-malicious factors, update frequency of the detection model, as well as storage and process of massive metering data (Singh et al. 2019; Jokar et al. 2016).

Smart Meter Management

Smart meter management regards the remote management and operation of SMs based on parameterizations, firmware upgrades and clock synchronization (Toledo 2013). Various information is set and can be updated during the meter operation, e.g., vendor, type, customer, location, age, tariff, and configuration settings, as well as meter status (e.g., battery condition, credit/prepayment mode), working life and records of safety and security checks. These data are stored and maintained in the meter database, while basic information of each device is integrated into the GIS for more efficient asset management and dispatch of field personnel.

SMs can detect meter faults and installation errors through automated on-event and on-demand diagnostic notification so that they can be addressed immediately rather than waiting for the next manual meter reading (S3C 2011). SM information, e.g., diagnostics and alarms, are integrated with workforce management systems and are included on work orders to allow field personnel to address premise-related issues more efficiently, ensuring minimum finance or service-related losses (Toledo 2013; KEMA International B.V 2012). Additional benefits for the energy providers include the simplifications in meter-to-cash process and customer change procedures, as well as the on-time accomplishment of firmware updates and calibration actions (Goel et al. 2011). Moreover, SM management can allow a quick supplier switch, an easier switch between billing schemes or load limits for the end-users (Waisi and Agyeman 2018). SMs are also used to set the switching times and season in case of multiple pricing registers, maximum demand registers or other registers related to electricity billing (Segovia and Sánchez 2011; DECC 2013).

The main requirement for the realization of SM management is a reliable and secure network for bidirectional communication between the meter and the utility, remote firmware updates and configuration of device settings (Toledo 2013; Goel et al. 2011). Alarm messages are also required to assist in diagnosis of faults, determination or confirmation of the meter state and investigation of suspicious events. Besides the update of tariff rates, switching times and seasons used for electricity billing, remote configuration is also required for the clock synchronization of SMs with the existing communication network by adopting the appropriate precision time protocol (Toledo 2013). Most SMs also have a demand limiting mode, which penalizes end-customers either by temporary disconnection, or by enforcing a higher electricity price, if a predefined demand threshold is exceeded. SM vendors or utilities also need to regularly update the firmware, which runs on the smart meter to control, monitor, and manipulate the data (Toledo 2013).

While major communications technologies for smart metering applications include radio frequency, PLC and 2G/3G/4G, the vast majority of SM communication networks are based on a combination of technologies (World Bank Group 2018b). In addition, lack of standardization of smart metering technologies has resulted in smart metering data management under different communication protocols (Rahman and Mto 2013). Another issue is that utilities usually apply communications protocols that rely on a specific spectrum band that might be unlicensed, since licensed spectrums have to be purchased for exclusive access (INHEMETER 2018). Since SMI components use traditional unlicensed bands, grid reliability and security might be jeopardized by cyber-attacks, causing significant data protection issues.

3.2 Customer-Oriented Smart Metering Applications

In this section, the smart metering applications that are of main interest for the end-customers, are presented.

Transparency

Customers equipped with conventional Ferraris meters can either manually or through meter readers collect and transfer only one meter reading to the supplier per billing period, which could be from once per year to monthly depending on the country (EC 2018). Since the consumption measurement is provided by either the bill or the manual reading, it is clear that the transparency of consumption with electromechanical meters is limited. Energy prices across the world are rising, thus, there is an important need for more frequent electricity metering in order to increase transparency in energy demand and to control the electrical loads more efficiently.

One of the fundamental SMI functionalities is to make energy consumption and energy costs more transparent to end-customers aiming at improved energy management. In particular, the SMs provide more information than a conventional meter and transfer the readings at regular intervals, e.g., 15-min, to the energy supplier (Tounquet 2019). Owing to the improved knowledge about the consumption patterns and the electricity price signals, the end-users can reduce or shift their demand leading to reduced consumption and improved economic welfare (Bollen 2011). Besides the SMs, the SMI can include additional IT applications, e.g., web portals, which increase awareness of energy usage, by visualizing load trends or using self-selected budgets. The customers will also benefit from more accurate invoices, easier switching procedures, more plausible energy bills, leading to greater confidence to their suppliers (KEMA International B.V 2012). Transparency for the end-users can also enable personalized advice with social experiences on a single device level (Akselrad et al. 2011).

Transparency of electricity consumption at end-user level is provided by indirect or direct feedback platforms (World Energy Council 2018; Darby 2006). Indirect feedback can be derived from utility data, including monthly or more frequent bills, while data can be processed by a third party to provide personally and socially contextual feedback. In terms of direct feedback, information technologies, such as IHDs and web-based information portals, are needed for the real-time visualization of SM data (Martinez et al. 2010; Armel et al. 2013). The SM gateway is also required for the data exchange among the devices and the utility or any third-parties (Alahakoon and Yu 2016). The network architecture, communication technology, as well as the capacity and speed of data transfer, shall ensure continuous visualization of energy consumption to the users (Karlin et al. 2014). Energy monitoring systems can include additional sensors when disaggregation of total electricity consumption into device level is applied (Akselrad et al. 2011; Karlin et al. 2014). In term of privacy issues, protecting data access through strict regulation is also required (Yang et al. 2019).

Near real-time monitoring of energy consumption is still feasible for the end-users in a small number of European countries, therefore, limiting the implementation of direct feedback platforms. High transparency through smart metering systems may require additional devices, e.g., sensors and meters, which may significantly increase the total costs arising questions about the real value proposition of the additional transparency (Yang et al. 2019). Finally, SMI gives rise to complex personal data processing operations, while most customers will be unaware of the nature of these operations and of the potential impact this could have on their privacy.

Improvement of Energy Efficiency

While the 2012 EU Energy Efficiency Directive established the energy efficiency target of 20% by 2020, the EC proposed an update to the Energy Efficiency Directive, including a new energy efficiency goal of 30% by 2030 (EU 2018). On the contrary to demand response (DR) methods that focus on reducing peak load during specific periods of high wholesale energy market prices, the energy efficiency measures are more comprehensive and focus on cost-effective energy consumption reductions in overall (Martinez et al. 2010).

A range of theory and research findings relate to how, and to what extent, customers can improve their energy efficiency, as a result of being equipped with smart metering systems. Various types of energy consumption feedbacks impose different levels of energy saving impacts on the consumer behavior (KEMA International B.V 2012). Studies show that the most successful energy efficiency measures involve both direct and indirect feedback from energy suppliers. The main benefits from the provision of information on individual electricity consumption are the change of consumer behavior, energy savings and active engagement in energy efficiency activities (KEMA International B.V 2012; Sachar et al. 2019; Popock et al. 2015). The end-users can develop a greater awareness of their electricity consumption, as they can notice which behaviors lead to high consumption (KEMA International

B.V 2012; Sachar et al. 2019). SMs can also provide information about equipment that need be replaced to increase the building energy efficiency (Sachar et al. 2019; Fischer 2008).

Energy savings with smart metering systems are highly dependent on the effectiveness of the feedback on energy use given to consumers, as well as the willingness and ability of the consumers to respond to this feedback. Furthermore, sophisticated behavioral nudging techniques and consumer interface design features are required to increase the impact of SMs (Sachar et al. 2019; BEIS 2017). Hence, the main prerequisites for the improvement of energy efficiency at consumers' side involve the presence of an appropriate energy information system, effective monetary and non-monetary incentives for the end-users to change their behaviors and a sufficient consumer education level (KEMA International B.V 2012; Sachar et al. 2019; BEIS 2017).

The main barrier for the engagement of end-customers in energy efficiency programs is the low motivation due to the lack of energy literacy and knowledge about the saving potential of different behavior changes (Popock et al. 2015; BEIS 2017). Consumers are usually unaware of appropriate behaviors to reduce their energy consumption, while a high motivation is insufficient to achieve considerable savings, since these actions need to be performed on long-term basis (BEIS 2017; Nachreiner et al. 2015). Furthermore, the customers are not willing to invest in energy efficiency measures, which cannot result in monetary savings in the near future. A significant proportion of daily energy consumption is also based on habitual cognitive or affective responses that the end-customers are demotivated to change. In summary, limited awareness and particular physical circumstances may be holding the end-users back from actively changing their energy consumption behavior.

End-Use Energy Management

End-use energy management is the key to saving energy for the end-users of all types, and it includes the process of monitoring, controlling and conserving energy. As a result, energy costs can be minimized while fulfilling the energy end-use requirements, e.g., keeping indoor conditions (temperature, air quality and lighting) or other processes within a certain range that also depends on the usage of the building. This is also related to the improvement of energy efficiency and the whole energy supply chains (Kádár and Varga 2012).

Energy management mainly depends on user awareness on energy consumption, as well as on the availability of relevant information (Zhou et al. 2010; Amaral et al. 2014). With SMs the energy consumption can be measured and analyzed with the main goal to find alternatives that can increase the energy savings and the RES self-consumption for prosumers. Customer benefits arise from the engagement in energy management by enabling information feedback about the energy consumption and the price signals (Armel et al. 2013). This information can lead to consumption reduction or load shifting and hence, to lower bills or mitigated cost increases. Consequently, the carbon footprint of utilities and end-users can also be reduced. Additional financial incentives can be provided to customers by accepting conservation measures when peak load across the grid becomes relevant (Yuan et al. 2020). Energy suppliers will also benefit from energy management, since congestion issues can be reduced, and power quality can be improved owing to better grid management (Ramchurn et al. 2011).

Essential requirement for the optimal energy management is the detailed interval energy consumption data, since patterns of energy waste can be better observed. Moreover, the disaggregation of each appliance used by the dwellings requires advanced processing and high sampling rates (Kong et al. 2018). In practice, energy management systems require near real-time consumption measurements preferably directly from the billing meters. Additional exploitation of smart meter potential can be achieved with smart plugs and energy management-enabled sensors, e.g., smart thermostats, while the interaction of customers with energy providers can considerably be improved. Energy management should also not jeopardize the operation of controllable devices, while the customers should not experience significant influence on their comfort level.

Various barriers related to social, technological, legal and economical aspects have delayed the deployment of energy management systems (Martinez et al. 2010; Karlin et al. 2014; Vasak et al. 2018; Zipperer et al. 2013; Yildiz et al. 2017). The main obstacle probably regards the investment in additional hardware and software to obtain near real-time feedback on the electricity consumption, leading to the uncertain cost effectiveness of energy management systems (Martinez et al. 2010; Bediako 2014). Another issue that exists in various European countries is the difficulty to monitor the energy consumption in near real-time due to data privacy and security concerns (Karlin et al. 2014; Vasak et al. 2018; Yildiz et al. 2017). Moreover, high sensitive and private information about the end-users may be used by third parties, thus, the exchange of data raises privacy and security issues (Karlin et al. 2014; Vasak et al. 2018; Yildiz et al. 2017). Lack of industry-accepted device communication and interoperability standards is also a critical issue to the widespread adoption of energy management systems (Vasak et al. 2018; Zipperer et al. 2013). Finally, lack of consumer awareness and knowledge is also cited as a barrier to the adoption of energy management systems (Karlin et al. 2014; Bediako 2014).

Smart Buildings

Smart buildings are defined as a set of communication technologies enabling different objects, sensors, and functions to communicate and interact with each other and also to be managed, controlled, and automated remotely (EC 2017). Core issues for the smart buildings are the energy management, indoor comfort management and building automation (Jia et al. 2019). In the literature, smart homes are also reported as a specific category of smart buildings, which also consists of several network-connected systems that can be remotely controlled or automated (4E EDNA 2018).

Smart homes describe the implementation of digitization and cross-linking in private residencies, while smart buildings involve the automation and centralized operation of technical equipment for big facilities.

The primary objectives of a smart building are to increase automation, facilitate energy management, enhance user comfort, and reduce environmental emissions (Groote et al. 2017). The smart buildings can benefit stakeholders on both sides of the interface—consumers, utilities and third parties—as there are strong incentives for all sides to help the others' function smoothly (Zipperer et al. 2013). SMs are important components of smart buildings, since they communicate with the energy suppliers, who possess an intelligent management service that can provide behavior-based feedback to the SMs for an efficient energy usage (Bhati et al. 2017). SM data are necessary for smart homes to enhance energy savings and to apply home automation and DR measures (Paetz et al. 2012). Other services provided by SMs in the context of smart home are safety and security by alarm signals, telemedicine by transmitting medical data and social alarms for elderly people (Koponen et al. 2008; Balmert and Petrov 2010).

One of the main requirements for the operation of smart buildings is to ensure the interoperability between SMs and the building appliances (4E EDNA 2018). Personalized, tailored and near real-time information and feedback on energy use and tariffs through SMs and IHDs are also required to enable user-control and programmed optimization of appliance use and any present micro-generation (Paetz et al. 2012). Besides that, access to near real-time data shall be given to authorized third parties, such as aggregators, installers, and energy brokers for the development of smart business solutions (Groote et al. 2017). In case of large-scale facilities, smart metering data can require the use of advanced data management platforms and interfaces for their analysis and interpretation (Note 2014).

Although the potential benefits of smart buildings are significant, energy users are still subject to several hurdles and concerns that could inhibit a rapid adoption of smart building technologies. Data privacy and security, as well as the potential for utilities to monitor or even control building demand have led to consumer backlashes against SMs (NEEP 2015; Rich et al. 2013). The provision of detailed energy use information to third-party vendors raises regulatory and legal issues that state legislatures and utility commissions must address. Moreover, interoperability of different smart appliances and their high up-front cost can also deter their adoption (Serrenho and Bertoldi 2019; Wilson et al. 2017). Uncertainties about the degree to which people have to change their lifestyles in order to save money also demotivate the end-users to invest in smart homes (Paetz et al. 2012). Other barriers include the complexity, technology risk, and lack of awareness about the capabilities and benefits of smart buildings (4E EDNA 2018; Rich et al. 2013).

3.3 Other Stakeholder-Oriented Smart Metering Applications

In this section, the smart metering applications that are of main interest for the authorities, institutions and associations are presented.

Provision of Statistical Data to Authorities

Smart metering data are valuable to various authorities, e.g., policy makers, regulation and legislation bodies seeking to understand and manage the energy use of customers (Elam 2016). The use of smart metering data is also expected to help the authorities develop suitable governmental policies and regulatory frameworks in the upcoming future.

SM data can also be utilized by authorities of official statistics for the development of national data hubs, which can be used by various stakeholders (Kesküla et al. 2010). In particular, such data hubs can provide energy statistics of businesses and households, and statistics on vacant living places or seasonal/temporary occupancy of living places. Regulators can also use the SMs for a better understanding of energy use and to improve the existing arrangements (Balmert and Petrov 2010). Moreover, SM data can be used by authorities to produce statistics on the reliability indices e.g., SAIDI, of distribution networks. Available SM data can help the authorities develop appropriate standards for both the hardware and software components of SMI. As for the policy-makers, SM data can also enhance evaluation of current or previous policies, while facilitating the development of future evidence-based policies in various areas, e.g., energy efficiency and fuel poverty (Elam 2016).

In order to maximize benefits from smart metering applications, the tariff schemes, technical codes, and procedures need to be adjusted to smart metering by regulators and policy-makers. Moreover, compliance of SMI with core principles of law, such as data minimization, legitimate data processing, transfer only for specific purposes, and data storage has to be considered by legislation bodies (Knyrim and Trieb 2011). Interoperability of smart metering systems and support of new services, e.g., DR and energy efficiency, should also be adopted by legislators, regulators and policy-makers when transposing the EU Directives and the Third Energy Package into specific national laws (Tounqouet 2019). Owing to the digitalization of energy sector, energy regulators have to cooperate with authorities from other sectors, such as data protection agencies, telecommunication regulators and monitoring price comparison tools, e.g., competition authorities (BEUC 2019).

While in several countries the grid operators are realizing the deployment of smart metering systems, the complete legal package for SMs has not been adopted by the national authorities yet (Tounqouet 2019; Vogt 2020). In particular, specific, and detailed requirements with regards to interoperability and near real-time measurement need to be adopted by the legislators (Vogt 2020). In addition, current EU legislation and communication consider a straightforward relationship between the customers, meters, and utility, which cannot be applied in practice for centralized energy departments that manage large and diverse portfolio of buildings with their own energy contracts (Vogt 2020). Another issue, which concerns the regulatory and legislation bodies, is the development of proper privacy policies and data security standards to ensure that customer energy consumption data are not accessed by unauthorized parties or misused (Balmert and Petrov 2010).

Provision of Data to R&D

Smart metering data are also valuable to academic and research institutions, since their use can enhance R&D in various areas of power systems, such as, end-use energy management, customer analysis, load forecasting and power quality control (Elam 2016).

In the frame of energy data analytics, data scientists are jointly trained by universities and industry to analyze massive amounts of smart metering data (Hong et al. 2018). Data analysts cooperate with third-party vendors that focus on consumers' behavior and preferences in the context of services' provision. Data are also used by regulating authorities and research centers in analyzing the energy consumption and behavior at customer level (Ulbig et al. 2016). Several academic and research institutions have already included the SMs in their electrical engineering curricula and research programs (Sayed et al. 2019). According to Elam (2016) and Environment and Energy (Technical Report), SM data can allow researchers to evaluate load forecasting models, customer clusters, impacts of energy management systems on occupants' well-being and comfort, as well as various DR programs and impact of different factors on demand patterns.

SM data are usually collected by data aggregators, e.g., utility companies. The data aggregator engages in negotiation with researchers and data analysts that are allowed to access the consumers' data considering data privacy and security aspects, qualitative data analysis and consumers' willingness to provide their data to third-parties. Hence, data are not shared with third parties other than to accredited researchers for approved scientific research. In most cases, identifiable personal data are pseudonymized or anonymized and accessed within a secure environment. Though high-resolution SM data enable new research avenues, contextual data that affect energy consumption patterns or consumer responses are also required by researchers for a complete analysis (Webborn and Oreszczyn 2019). Such data are related to tariff schemes, socio-demographic attributes (e.g., family size and age), building characteristics or types of activities for non-residential consumers. As highlighted in Webborn and Oreszczyn (2019), contextual data can only be provided in a pseudonymized form due to consumers' data protection issues.

One of the main challenges faced by researchers in SM data analytics is the availability of datasets. Due to security and privacy issues, power industries are reluctant to release their measurement data for data analytics research purposes (Hu and Vasilakos 2016). Common barriers to high-quality research using SM data can also be the lack of raw and contextual data, data quality issues and low sample size (Webborn et al. 2019). Due to the privatization of energy data, coordinated efforts are needed from governments, funding bodies and researchers, as well as legislation or goodwill on the part of utility companies to overcome the barriers to data access (Webborn and Oreszczyn 2019).

4 Current Status of Smart Metering Applications

4.1 Questionnaire

A questionnaire was created to determine the extent to which these applications are known and used by Swiss businesses. Each application was first briefly discussed. Six more specific questions are asked in the case if the respondent is aware of the application:

1. "Do you think that the application is technically applicable?" (yes/no)
2. "Was that application already considered and analyzed in your company?" (yes/no)
3. "Is the application already implemented in your company?" (yes/no)
4. "What do you think about the stakeholder with the most interests?" (multiple choice: consumer, producer, seller, DSO, TSO, government, others)
5. "How do you estimate the potential of this application?" [technical: 1 (low)–4 (high), economical: 1 (low)–4 (high)]
6. "What are the barriers for the implementation of this application in Switzerland today?"

In December 2020, the questionnaire was sent to 18 stakeholders. To achieve a high return rate, the addressed people, who work for Swiss utilities, were chosen based on the authors' contacts. Thirteen of them, or 72% of the total number of candidates, responded anonymously. Table 4 shows the findings of the questionnaire, with the exception of the last question, which was answered using free-text solutions.

4.2 Results

All respondents are familiar with five applications, which may be the most commonly mentioned business cases in the context of SMs in Switzerland. However, at least half of the respondents are aware of the remaining applications. The low score for "Predictive Maintenance and Failure Analysis" stands out, probably because the Swiss grid is so reliable.

Almost every application is deemed technically feasible. The majority of applications, according to the respondents, can be executed in Switzerland. In addition, the majority of respondents thought about and analyzed all of the applications. However, just a few applications have already been adopted by respondents' companies.

Table 4 The findings of the questionnaire

	Known	Technically feasible	Considered and analyzed	Implemented	Technical potential	Economical potential
Settlement and billing	13 (100%)	13 (100%)	11 (85%)	7 (54%)	3.15	2.46
Smart meter management	13 (100%)	13 (100%)	10 (77%)	8 (62%)	3.08	2.31
Distribution system state estimation	13 (100%)	13 (100%)	10 (77%)	2 (15%)	3.08	2.54
Predictive maintenance and analysis of failures	7 (54%)	6 (46%)	5 (38%)	0 (0%)	2.43	2.00
Monitoring of power quality	8 (62%)	8 (62%)	6 (46%)	5 (38%)	2.75	2.50
Load modelling and forecasting	10 (77%)	10 (77%)	5 (38%)	2 (15%)	3.20	3.00
Customer analysis	9 (69%)	8 (62%)	6 (46%)	2 (15%)	2.78	2.44
Enhanced efficiency and competition in energy markets	9 (69%)	7 (54%)	5 (38%)	2 (15%)	2.78	2.54
Demand side management	12 (92%)	11 (85%)	9 (69%)	3 (23%)	3.00	2.67
Electrical theft detection	9 (69%)	9 (69%)	4 (31%)	2 (15%)	2.56	1.89
Transparency	13 (100%)	13 (100%)	8 (62%)	7 (54%)	2.62	1.77
Improvement of energy efficiency	13 (100%)	12 (92%)	6 (46%)	5 (38%)	2.15	1.92
End-use energy management	8 (62%)	7 (54%)	6 (46%)	1 (8%)	3.13	2.50
Smart buildings	9 (69%)	9 (69%)	4 (31%)	1 (8%)	2.22	2.00
Provision of statistical data to authorities	6 (46%)	6 (46%)	3 (23%)	2 (15%)	2.33	2.17
Provision of data to R&D centres	6 (46%)	6 (46%)	4 (31%)	2 (15%)	2.83	2.17

Based on Swiss law, which mandates that SMs be installed for 80% of all users by the end of 2027, the two most popular applications, "Settlement and Billing" and "Smart Meter Management," are examined by almost 80% of the respondents. However, applications such as "Distribution System State Estimation", "Predictive Maintenance and Failure Analysis", and "Monitoring of Power Quality," which ensure grid quality, were also assessed. However, only "Monitoring of Power Quality" has a high implementation rate, in contrast to the previous two programs, which are both used by about 60% of respondents. The other two, "Distribution System State Estimation" and "Predictive Maintenance and Failure Analysis," appear to be unimportant to the respondents' business. This is notably true for "Distribution System State Estimation," since all respondents are aware of it, ten have examined it, but only two have implemented it.

Other applications were evaluated by 58% of respondents' organizations on average, but just 20% of all respondents deployed them. This percentage could be read as indicating that these applications are not valuable to the companies of the respondents. The application "Transparency," for example, was studied and analyzed by 62% of respondents, and was previously installed by 54% already.

The low adoption rate of smart meter applications appears to be verified by the respondents' appraisal of their technical and economic potential. The average and standard deviation of the technical (orange) and economic (blue) potentials are shown in Fig. 5. It's worth noting that the economic potential for any application is smaller than the technical potential. The difference in technical and economic possibilities is particularly noticeable for "Settlement and Billing", "Smart Meter Management", and "Transparency."

The applications with the lowest economical potential are "Transparency" (1.77), "Electrical Theft Detection" (1.89), "Improvement of Energy Efficiency" (1.92), "Predictive Maintenance and Failure Analysis" (2), and "Smart Buildings" (2). Even the latter may come as a surprise, as various companies are involved in the field of "Smart Buildings." But the standard deviation shows that the assessment of the potential varies a lot (e.g., 1.05 for the economical potential of "Electrical Theft Detection"). There appears to be agreement only for "Load Modelling and Forecasting" (0.47), "Distribution System State Estimation" (0.52), and "Transparency" (0.6).

Out of all smart meter applications, only "Improvement of Energy Efficiency" (2.15), "Smart Buildings" (2.22), and "Provision of Statistical Data to Authorities" (2.33) have the lowest technical potential. "Settlement and Billing" (3.15), "End-Use Energy Management" (3.13), "Smart Meter Management" (3.08), and "Distribution System State Estimation" are expected to have the most potential (3.08). However, the standard deviation for the technical potential is higher than that for the economic potential, indicating that the respondents may not totally agree on the application's technical potential. They do, however, agree on the great potential of "Load Modelling and Forecasting" (0.63), which has a high agreed-upon economic potential. The agreement is rather high for "End-Use Energy Management" (0.64), as is the estimate of the technological potential itself, but the economical potential is averaged.

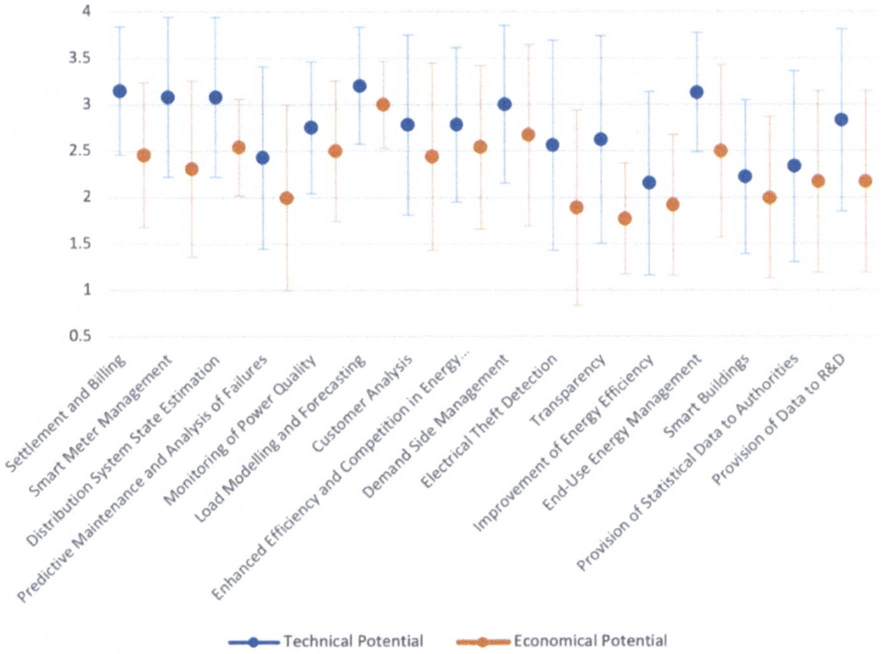

Fig. 5 Technological and economical potential of SM applications

Respondents could describe the difficulty of introducing such application in the last question. The regulatory framework and present law are two of the most prominent issues raised by the responders. The continual roll-out of smart meters and related applications is hampered by policy changes and several ambiguities. As a result, smart meters lack a defined (but proprietary) interface, which is required for some applications. Furthermore, smart meter certifications in accordance with current governmental criteria are lacking. The existing Swiss law restricts the complete implementation of various apps, which may be one of the reasons why so few are implemented. Furthermore, data privacy appears to be impeding the implementation of some smart meter applications, as data from smart meters can only be held for 12 months and read once per day. However, some applications necessitate larger data storage, higher retrieval speeds, and more data than only power consumption or output. One respondent even claimed that the most notable business case for smart meter adoption in Switzerland, "Settlement and Billing", which is supposed to save money, did not save money in the end. "Electrical Theft Detection", according to several respondents, is not a problem in Switzerland. Consumers are considered as a stakeholder in all smart meter applications, yet "99% of all consumers are uninterested in their energy consumption," according to the remarks in the last question. In addition, a smart meter is not required for every application. "End-Use Energy Management" and "Smart Buildings", for example, may be carried out more easily without the involvement of SMs, as some responders indicated. Some respondents reject the last two applications, "Provision of Statistical Data to Authorities" and "Provision of Data to R&D", because they would entail additional work for which they might not be compensated.

To summary, the questionnaire revealed some intriguing information. The respondents are aware of a variety of smart meter applications, but just a few have been implemented. Smart meters' poor economic potential, as well as worries about government agencies and policies, limit their use for applications other than the "typical" ones.

5 Discussion and Conclusion

As concluded by the questionnaire, smart meter management, settlement and billing, and transparency are the mostly known and implemented applications in Switzerland. Regarding the economical potential of the aforesaid applications in Switzerland, it was clear that transparency has the lowest one, while SM management and settlement and billing have average potential among all applications. This could additionally support the hypothesis that the SMI is not implemented owing to economical benefits, but due to other reasons, e.g., law and marketing. Similar conclusions were derived for the electrical theft detection, which is not an issue in Switzerland, therefore, it has low implementation rate and economical potential. Main issue to be encountered for the aforesaid applications is the frequent provision of SM data, while cyber-tampering is of main concern for SM management, transparency, and electricity theft detection.

In terms of the smart metering applications related to the grid status, e.g., DSSE, power quality monitoring, and predictive maintenance and analysis of failures, the energy service providers are mainly interested in their implementation. It was concluded that the need for near real-time measurements, incomplete or inaccurate data, as well as demanding capabilities for data storage infrastructure, information management programs and communication networks demotivate the utilities to invest in. Moreover, it is still unclear to which degree the DSOs could benefit from these applications taking into account the high deployment costs, as well as the secure and stable status of the current distribution networks. While the respondents to the questionnaire see high technical and economical potential for DSSE, power quality monitoring and predictive maintenance and analysis of failures are considered to have a moderate economical potential. Furthermore, only power quality monitoring is implemented by a number of the respondents, while the other two applications have lower implementation rates.

Concerning the applications of load modelling and forecasting, and customer analysis, the energy service providers are also mainly interested in their implementation. Main barriers for the wide applicability of the aforesaid applications are the current reading resolution of SMs, lack of ground truth data and bad or missing data. Moreover, the uncertain influence of various factors on load profiles, the high computational time and effort for the acquisition and organization of large-scale datasets, as well as the complexity of data-driven techniques and data privacy issues have delayed their wide implementation. From the questionnaire, it was concluded that load modelling and forecasting has higher technical and economical potential than customer analysis, however, both applications are only implemented by two respondents.

Regarding the DSM and enhanced efficiency and competition in energy markets, energy suppliers usually offer cost-effective contracts based on DSM programs to large-scale consumers, while aggregated flexibilities by a group of small customers are offered by BRPs in the energy markets. However, several issues, e.g., low monetary benefits and lack of awareness or trust from the customer side, stymie their engagement in DSM. Though the respondents to the questionnaire confirm the high DSM potential and most of them already considered DSM for their company, only 3 implemented it. Moreover, the application of enhanced efficiency and competition in energy markets had lower techno-economical potential, and implementation rate than DSM.

As for the end-customer oriented applications, improvement of energy efficiency, end-use energy management and smart buildings can provide significant benefits to various stakeholders. However, the value of energy savings is still unclear, and the high up-front costs demotivate the customers to be engaged due to the lack of attractive pricing schemes. The lack of consumer awareness and uncertainties about the degree to which end-users have to change their lifestyles are also main challenges. Additional barriers are related to the high complexity, uncertain performance, interoperability problems and data privacy issues. From the questionnaire results, only improvement of energy efficiency is implemented by several respondents, while end-use energy management and smart buildings were only implemented by one participant. Nevertheless, end-use energy management is already considered by most respondents, certifying the highest technical and economical potential of the aforesaid applications.

Apart from the energy service providers and the end-customers, SM data are valuable to various authorities, e.g., policy makers, regulation and legislation bodies, and R&D direction, e.g., associations, academic and research institutes, however, the complete legal package for SMs has not been fully clarified. In particular, specific, and detailed requirements with regards to interoperability and near real-time measurement need to be adopted by the legislators. Utilization of smart metering data by authorities and R&D may be hindered due to security and privacy issues, therefore, coordinated efforts are required by various stakeholders to overcome the barriers to data access. Provision of smart metering data to other stakeholders was known by the half of the questionnaire respondents, and one respondent also mentioned that the unnecessary bureaucracy may demotivate the implementation of the relevant applications.

References

4E EDNA (2018) Intelligent efficiency—a case study of barriers and solutions—smart homes. Technical report. 4E Electronic Devices & Networks Annex, pp 1–55

Abur A, Expósito AG (2004) Power system state estimation: theory and implementation. Marcel Dekker Inc., New York, Basel, pp 1–346

Ahmad F, Rasool A, Ozsoy E, Sekar R, Sabanovic A, Elitaş M (2018) Distribution system state estimation—a step towards smart grid. Renew Sustain Energy Rev 81:2659–2671

Akselrad D, Petcu V, Römer B, Schmid A, Bytschkow D, Engelken M (2011) Making home energy usage transparent for households using smart meters. In: Proceedings of IEEE international conference on consumer electronics (ICCE), Berlin, Sept 2011, pp 150–153

Alahakoon D, Yu X (2016) Smart electricity meter data intelligence for future energy systems: a survey. IEEE Trans Ind Inf 12(1):425–436

Albu MM, Sănduleac M, Stănescu C (2017) Syncretic use of smart meters for power quality monitoring in emerging networks. IEEE Trans Smart Grid 8(1):485–492

Ali S, Wu K, Weston K, Marinakis D (2016) A machine learning approach to meter placement for power quality estimation in smart grid. IEEE Trans Smart Grid 7(3):1552–1561

Alimardani A, Therrien F, Atanackovic D, Jatskevich J, Vaahedi E (2015) Distribution system state estimation based on nonsynchronized smart meters. IEEE Trans Smart Grid 6(6):2919–2928

Amaral HLMD, Souza AND, Gastaldello DS, Fernandes F, Vale Z (2014) Smart meters as a tool for energy efficiency. In: Proceedings of IEEE/IAS international conference on industry applications (EEE INDUSCON), Juiz de Fora, Dec 2014, pp 1–6

Arif A, Wang Z, Wang J, Mather B, Bashualdo H, Zhao D (2018) Load modeling—a review. IEEE Trans Smart Grid 9(6):5986–5999

Armel KC, Gupta A, Shrimali G, Albert A (2013) Is disaggregation the holy grail of energy efficiency? The case of electricity. Energy Policy 52:213–234

Asghar MR, Dán G, Miorandi D, Chlamtac I (2017) Smart meter data privacy: a survey. IEEE Commun Surv Tutorials 19(4):2820–2835

Atos Worldgrid (2010) Electricity smart metering business drivers. Technical report, pp 1–20

Balmert D, Petrov K (2010) Regulatory aspects of smart metering. ERRA Licensing and Competition Committee. Issue paper. KEMA, pp 1–72

Beckel C, Sadamori L, Santini S, Staake T (2015) Automated customer segmentation based on smart meter data with temperature and daylight sensitivity. In: Proceedings of IEEE international conference on smart grid communications (SmartGridComm), Miami, Nov 2015, pp 653–658

Bediako BA (2014) SMART energy homes and the smart grid: a framework for intelligent energy management systems for residential customers. Doctoral thesis. Technische Universiteit Eindhoven, pp 1–168

BEIS (2017) Smart metering energy efficiency advice project. Annex 1: review of energy efficiency advice best practice. Technical report. Department of Energy and Climate Change, pp 1–42

Berg FVD, Kadurek P, Cobben S, Kling W (2011) Electricity theft localization based on smart metering. In: Proceedings of 21st international conference on electricity distribution (CIRED), Frankfurt, June 2011, pp 1–4

BEUC (2019) The future of energy consumers—bright or burdensome? Technical report. BEUC—The European Consumer Organization, The Consumer Voice in Europe, pp 1–28

Bhati A, Hansen M, Chan CM (2017) Energy conservation through smart homes in a smart city: a lesson for Singapore households. Energy Policy 104:230–239

Bogdanovic M, Schütz T, Cupelli M, Copeland M, Olson JE (2020) InterFlex—investigation and comparison of EU-wide regulations and rules concerning the commercialization of end-customers flexibility and building local energy market places/platforms, V1.0, D8.9. Horizon 1–70

Bollen M (2011) Adapting electricity networks to a sustainable energy system—smart metering and smart grids. Technical report. Energy Markets Inspectorate, pp 1–115

Borges FAS, Fernandes RAS, Silva IN, Silva CBS (2016) Feature extraction and power quality disturbances classification using smart meters signals. IEEE Trans Ind Inf 12(2):824–833

Campbell M, Watson N, Miller A (2015) Smart meters to monitor power quality at consumer premises. In: Proceedings of Electricity Engineers Association conference (EEA), Wellington, June 2015, pp 1–12

Carroll P, Murphy T, Hanley M, Dempsey D, Dunne J (2018) Household classification using smart meter data. J Official Stat 34(1):1–25

CEC (2014) Reforming the Energy Vision (REV) Working Group I: customer engagement. Staff report on the work of the Customer Engagement Committee, pp 1–224

CEI/IEC (2003) 61000-4-30:2003, International standard "Electromagnetic compatibility (EMC)—part 4-30: testing and measurement techniques—power quality measurement methods. Standard, International Electrotechnical Commission, pp 1–98

CENELEC (2010) EN 50160 voltage characteristics of electricity supplied by public electricity networks. Standard, European Committee for Electrotechnical Standardization, pp 1–34

Cerquitelli T, Chicco G, Corso ED, Ventura F, Montesano G, Pizzo AD, González AM, Sobrino EM (2018) Discovering electricity consumption over time for residential consumers through cluster analysis. In: Proceedings of 14th international conference on development and application systems (DAS), Suceava, May 2018, pp 164–169

Chang H, Huang Y, Ebrahimi S, Jatskevich J (2017) Smart meter based selective harmonics compensation in buildings distribution systems with AC/DC microgrids. In: Proceedings of IEEE power and energy society general meeting (PESGM), Chicago, July 2017, pp 1–5

Chaouch M (2014) Clustering-based improvement of nonparametric functional time series forecasting: application to intra-day household-level load curves. IEEE Trans Smart Grid 5(1):411–419

Chatterjee N, Glick R, McNamee B (2019) Assessment of demand response and advanced metering. Staff report. Federal Energy Regulatory Commission, pp 1–44

Chitsaz H, Shaker H, Zareipour H, Wood D, Amjady N (2015) Short-term electricity load forecasting of buildings in microgrids. Energy Build 99:50–60

Chuan L, Ukil A (2015) Modeling and validation of electrical load profiling in residential buildings in Singapore. IEEE Trans Power Syst 30(5):2800–2809

CMA (2015) Energy market investigation—gas and electricity settlement and metering. Technical report. Competition & Markets Authority, UK, pp 1–31

Conchado A, Linares P (2012) The economic impact of demand-response programs on power systems. A survey of the state of the art. Working paper 02-2010. Economics for Energy, pp 1–23

Czechowski R, Kosek AM (2016) The most frequent energy theft techniques and hazards in present power energy consumption. In: Proceedings of joint works on cyber-physical security and resilience in smart grids (CPSR-SG), Vienna, April 2016, pp 1–7

Darby S (2006) The effectiveness of feedback on energy consumption—a review for DEFRA of the literature on metering, billing and direct displays. Environmental Change Institute, University of Oxford, pp 1–24

Day P, Fabian M, Noble D, Ruwisch G, Spencer R, Stevenson J, Thoppay R (2014) Residential power load forecasting. Procedia Comput Sci 28:457–464

DECC (2013) Smart metering implementation programme—smart metering equipment technical specifications, ver. 2. Department of Energy and Climate Change, pp 1–92

Der Schweizerische Bundesrat (2013) Botschaft zum ersten Massnahmenpaket der Energiestrategie 2050 und zur Volksinitiative «Für den geordneten Ausstieg aus der Atomenergie (Atomausstiegsinitiative)». Swiss Confederation, pp 7561–7756

DOE (2006) Benefits of demand response in electricity markets and recommendations for achieving them. A report to the United States Congress Pursuant to section 1252 of the Energy Policy Act of 2005. U.S. Department of Energy, pp 1–97

EC (2016) Second consumer market study on the functioning of the retail electricity markets for consumers in the EU. Final report. European Commission, pp 1–394

EC (2017) Smart building: energy efficiency application. Digital Transformation Monitor, pp 1–6
EC (2018) Consumer study on "Pre-contractual information and billing in the energy market - improved clarity and comparability." Technical report. European Commission, Publications Office, pp 1–218
Efkarpidis N, Geidl M, Wache H, Peter M, Adam M (2022) Smart metering applications: main concepts and business models. In: Springer lecture notes in energy. Springer Nature Switzerland, Basel, pp 1–164
Eibl G, Bao K, Grassal PW, Bernau D, Schmeck H (2018) The influence of differential privacy on short term electric load forecasting. Energy Inf 1(48):93–113
Eissa MM (2011) Demand side management program evaluation based on industrial and commercial field data. Energy Policy 39(10):5961–5969
Elam S (2016) Smart meter data and public interest issues—the national perspective. Discussion paper 1, pp 1–41
ELEXON (2010) Profiling and settlement review—supplier consultation. Executive Summary, Consumer Focus and Accenture, pp 1–27
ENA (2012) Analysis of network benefits from smart meter message flows. Technical report. Energy Networks Association, pp 1–32
Environment and Energy. Using smart meter data to enable energy demand research—data service as a platform. Technical report, pp 1–2
EnWG (2005) Gesetz über die Elektrizitäts- und Gasversorgung (Energiewirtschaftsgesetz - EnWG). Bundesministerium für Wirtschaft und Klimaschutz, pp 1–114
EPRI (2011) End-use load composition estimation using smart meter data. Technical report. Electric Power Research Institute, pp 1–90
ERGEG (2007) Smart metering with a focus on electricity regulation. Technical report. European Regulators' Group for Electricity and Gas, pp 1–62
ERGEG (2009) Status review on regulatory aspects of smart metering (electricity and gas) as of May 2009. Technical report. European Regulators' Group for Electricity and Gas, pp 1–69
EU (2009) Directive 2009/72/EC of the European Parliament and of the Council of 13 July 2009 concerning common rules for the internal market in electricity and repealing Directive 2003/54/EC. Official J Eur Union 211:55–93
EU (2018) Directive (EU) 2018/2002 of the European Parliament and of the Council of 11 December 2018 amending Directive 2012/27/EU on energy efficiency. Official J Eur Union 328:210–230
EU (2019) Directive (EU) 2019/944 of the European Parliament and of the Council of 5 June 2019 on common rules for the internal market for electricity and amending directive 2012/27/EU. Official J Eur Union 158:125–199
EURELECTRIC (2017) Dynamic pricing in electricity supply. A EURELECTRIC position paper, pp 1–16
European Smart Grids Task Force Expert Group 3 (2019) Demand side flexibility—perceived barriers and proposed recommendations. Final report, pp 1–50
Figueiredo V, Rodrigues F, Vale Z, Gouveia JB (2005) An electric energy consumer characterization framework based on data mining techniques. IEEE Trans Power Syst 20(2):596–602
Fischer C (2008) Feedback on household electricity consumption: a tool for saving energy? Energ Effi 1(1):79–104
Flath C, Nicolay D, Conte T, Dinther CV, Neumann LF (2012) Cluster analysis of smart metering data: an implementation in practice. Bus Inf Syst Eng 4(1):31–39
Gellings CW (1985) The concept of demand-side management for electric utilities. Proc IEEE 73(10):1468–1470
Goel S, Hong Y, Papakonstantinou V, Kloza D (2011) Smart grid security. In: Springer briefs in cybersecurity. Springer, London, pp 1–129
Grigoras G, Ivanov O, Gavrilas M (2014) Customer classification and load profiling using data from smart meters. In: Proceedings of 12th symposium on neural network applications in electrical engineering (NEUREL), Belgrade, Nov 2014, pp 1–5

Groote MD, Volt J, Bean F (2017) Is Europe ready for the smart buildings revolution? Mapping smart-readiness and innovative case studies. Technical report. Buildings Performance Institute Europe (BPIE), pp 1–36

Haben S, Singleton C, Grindrod P (2016) Analysis and clustering of residential customers energy behavioral demand using smart meter data. IEEE Trans Smart Grid 7(1):136–144

Hejazi HA, Rad HM (2018) Power systems big data analytics: an assessment of paradigm shift barriers and prospects. Energy Rep 4:91–100

Herbst I, Lukovic S, Gasparin A, Schulz N, Witzig J, Kieber S (2019) LV grid data analysis demonstrated at DSO arbon energie. In: Proceedings of 25th international conference on electricity distribution (CIRED), Madrid, June 2019, pp 1–5

Hong T, Gao DW, Laing T, Kruchten D, Calzada J (2018) Training energy data scientists: universities and industry need to work together to bridge the talent gap. IEEE Power Energy Mag 16(3):66–73

Hu J, Vasilakos AV (2016) Energy big data analytics and security: challenges and opportunities. IEEE Trans Smart Grid 7(5):2432–2436

IBM (2012) Managing big data for smart grids and smart meters. White paper. Information management, pp 1–8

INHEMETER (2018) Metering and smart energy international. Technical report, Issue 3, pp 1–80

Jakaria AHM, Rahman MA, Hasan MGMM (2019) Safety analysis of AMI networks through smart fraud detection. In: Proceedings of IEEE conference on communications and network security (CNS), Washington, June 2019, pp 1–7

Jawurek M, Johns M, Kerschbaum F (2011) Plug-in privacy for smart metering billing. Lecture notes in computer science, vol 6794, pp 192–210

Jia M, Komeily A, Wang Y, Srinivasan RS (2019) Adopting internet of things for the development of smart buildings: a review of enabling technologies and applications. Autom Constr 101:111–126

Jokar P, Arianpoo N, Leung VCM (2016) Electricity theft detection in AMI using customers' consumption patterns. IEEE Trans Smart Grid 7(1):216–226

Jonhson L, Potreck LL, Seifert P, Nagao S (2018) Getting smart about smart meter analytics. White paper. Siemens EnergyIP Analytics, pp 1–20

Kádár P, Varga A (2012) The role of the smart meters in the energy management systems. IFAC Proc 45(21):121–125

Kadurek P, Blom J, Cobben JFG, Kling WL (2010) Theft detection and smart metering practices and expectations in the Netherlands. In: Proceedings of IEEE PES innovative smart grid technologies conference Europe (ISGT), Gothenburg, Oct 2010, pp 1–6

Karlin B, Ford R, Squiers C (2014) Energy feedback technology: a review and taxonomy of products and platforms. Energ Effi 7(3):377–399

KEMA International B.V (2012) Development of best practice recommendations for smart meters rollout in the energy community. Technical report. Energy Community, pp 1–107

Kemal M, Sanchez R, Olsen R, Iov F, Schwefel HP (2020) On the trade-off between timeliness and accuracy for low voltage distribution system grid monitoring utilizing smart meter data. IEEE Trans Power Syst 121

Kesküla A, Raitviir T, Jansson I, Pekarskaya T, Fosen J, Holm MR (2010) Implementation of smart meter data in the production of official statistics. ESSnet big data II—WP D smart energy, deliverable 3, pp 1–58

Khan ZA, Jayaweera D (2019) Smart meter data based load forecasting and demand side management in distribution networks with embedded PV systems. IEEE Access 8:2631–2644

Khan ZA, Jayaweera D, Alvarado MSA (2018) A novel approach for load profiling in smart power grids using smart meter data. Electr Power Syst Res 165:191–198

Khuntia SR, Rueda JL, Bouwman S, Meijden MAMMVD (2016) A literature survey on asset management in electrical power [transmission and distribution] system. Int Trans Electr Energy Syst 26:2123–2133

Kim Y, Aravkin A, Fei H, Zondervan A, Wolf M (2016) Analytics for understanding customer behavior in the energy and utility industry. IBM J Res Dev 60(1):1–13

Kinetica (2018) Solving the extreme data challenge for utilities. White paper, pp 1–17

Knyrim R, Trieb G (2011) Smart metering under EU data protection law. Int Data Priv Law 1(2):121–128

Kong W, Dong ZY, Ma J, Hill DJ, Zhao J, Luo F (2018) An extensible approach for non-intrusive load disaggregation with smart meter data. IEEE Trans Smart Grid 9(4):3362–3372

Koponen P, Saco LD, Orchard N, Vorisek T, Parsons J, Rochas C, Morch AZ, Lopes V, Togeby M (2008) Definition of smart metering and applications and identification of benefits. Technical report. WP2, Del. 3, ver. 1.1. ESMA, pp 1–42

Kuzkina Y, Golub I (2018) Smart meters as a key component of modern measuring infrastructure providing observability and state estimation of low-voltage distribution networks. In: E3S web of conferences, vol 69, pp 1–6

Kwac J, Tan CW, Sintov N, Flora J, Rajagopal R (2013) Utility customer segmentation based on smart meter data: empirical study. In: Proceedings of IEEE international conference on smart grid communications (SmartGridComm), Vancouver, Oct 2013, pp 720–725

Kwac J, Flora J, Rajagopal R (2014) Household energy consumption segmentation using hourly data. IEEE Trans Smart Grid 5(1):420–430

Landis+Gyr (2014) A guide to smart metering. Technical report. Landis+Gyr AG and Serus Media Oy, pp 1–14

Landis+Gyr (2021) Manage energy better together. Landis+Gyr, pp 1–49

Li P, Zhang B, Weng Y, Rajagopal R (2017) A sparse linear model and significance test for individual consumption prediction. IEEE Trans Power Syst 32(6):4489–4500

Liao H, Milanović JV (2016) Pathway to cost-efficient state estimation of future distribution networks. In: 2016 IEEE power and energy society general meeting (PESGM), pp 1–5

Maheswaran D, Selvaraj V, Manjaly DP (2015) Power quality monitoring systems for future smart grids. In: Proceedings of 23rd international conference on electricity distribution (CIRED), Lyon, June 2015, pp 1–5

Martinez KE, Donnelly KA, Laitner JA, York D, Talbot J, Friedrich K (2010) Advanced metering initiatives and residential feedback programs: a meta-review for household electricity-saving opportunities. American Council for an Energy-Efficient Economy, pp 1–128

McLaughlin S, Holbert B, Fawaz A, Berthier R, Zonouz S (2013) A multi-sensor energy theft detection framework for advanced metering infrastructures. IEEE J Sel Areas Commun 31(7):1319–1330

McLoughlin F, Duffy A, Conlon M (2012) Characterising domestic electricity consumption patterns by dwelling and occupant socio-economic variables: an Irish case study. Energy Build 48:240–248

Messinis GM, Rigas AE, Hatziargyriou ND (2013) A hybrid method for non-technical loss detection in smart distribution grids. IEEE Trans Smart Grid 10(6):6080–6091

Moghaddass R, Wang J (2018) A hierarchical framework for smart grid anomaly detection using large-scale smart meter data. IEEE Trans Smart Grid 9(6):5820–5830

Mohassel RR, Fung A, Mohammadi F, Raahemifar K (2014) A survey on advanced metering infrastructure. Int J Electr Power Energy Syst 63:473–484

Mora R, Elmer J, Venkatesan P, Seitz S, Jeston S, Zeug B, Mahurkar-Thombre J (2019) Capital markets day. Landis+Gyr, pp 1–104

Morch AZ, Parsons J, Ketser JCP (2007) Smart electricity metering as an energy efficiency instrument: comparative analyses of regulation and market conditions in Europe. In: Proceedings of ECEEE 2007 summer study on energy efficiency: saving energy—just do it!, Cote d'Azur, pp 142–149

Moreira J, Matos R, Guerra F, Paulo PS, Nunes PM, Matos PG, Pires R, Guisado J, Simões T, Pires G, Santos M, Pereira P, Felício P, Ahmad Y, Alcarva M, Chamorrinha V, Mousinho P (2020) Upgrid WP4—demonstration in real user environment: EDPD—Portugal, D4.3—evaluation of demonstration results and data collection. Horizon 1–165

Music M, Bosovic A, Hasanspahic N, Avdakovic S, Becirovic E (2013) Integrated power quality monitoring system and the benefits of integrating smart meters. In: Proceedings of 8th international conference on compatibility and power electronics (CPE), Ljubljana, June 2013, pp 86–91

Nachreiner M, Mack B, Matthies E, Mai KT (2015) An analysis of smart metering information systems: a psychological model of self-regulated behavioural change. Energy Res Soc Sci 9:85–97

NEEP (2015) Opportunities for home energy management systems (HEMS) in advancing residential energy efficiency programs. HEMS research report. Northeast Energy Efficiency Partnerships, pp 1–110

Noh HY, Rajagopal R (2013) Data-driven forecasting algorithms for building energy consumption. In: Proceedings SPIE 8692, sensors and smart structures technologies for civil, mechanical, and aerospace systems, San Diego, April 2013, pp 1–8

Northeast Group LCC (2017) $96 billion is lost every year to electricity theft. Available Online: http://www.northeast-group.com

Nursimulu A (2015) Demand-side flexibility for energy transitions—ensuring the competitive development of demand response options. Technical report. International Risk Governance Council (IRGC), pp 1–53

Obenchain GT, Thurber J, Queen EE, Gilleland H, Holland L, Hawkins A, Bender K, Morgan T, Barto L (2011) Smart meters and smart meter systems: a metering industry perspective. An EEI-AEIC-UTC white paper. EEI, AEIC, UTC, pp 1–35

Osterwalder A, Pigneur Y (2010) Business model generation. Wiley, Hoboken, pp 1–288

Otuoze AO, Mustafa MW, Mohammed OO, Saeed MS, Bakinde NTS, Salisu S (2019) Electricity theft detection by sources of threats for smart city planning. IET Smart Cities 1(2):52–60

Owen G, Ward J (2008) The consumer implications of smart meters. Technical report. Sustainability First, National Consumer Council, pp 1–45

Paetz AG, Dütschke E, Fichtner W (2012) Smart homes as a means to sustainable energy consumption: a study of consumer perceptions. J Consum Policy 35(1):23–41

Parag Y, Sovacool BK (2016) Electricity market design for the prosumer era. Nat Energy 1(4):1–6

Parvez I, Aghili M, Sarwat AI, Rahman S, Alam F (2019) Online power quality disturbance detection by support vector machine in smart meter. J Mod Power Syst Clean Energy 7(5):1328–1339

Pau M (2020) FLEXMETER—flexible smart metering for multiple energy vectors with active prosumers. Report on evaluation against defined metrics and scaling issues. Horizon 1–86

Peppanen J, Reno MJ, Thakkar M, Grijalva S, Harley RG (2015) Leveraging AMI data for distribution system model calibration and situational awareness. IEEE Trans Smart Grid 6(4):2050–2059

Pokhrel BR, Jensen BB, Pillai JR (2019) Integrated approach for network observability and state estimation in active distribution grid. Energies 12(11):1–17

Polgári B, Raisz D (2012) Application of smart meters especially for the detection of illegal electricity usage. In: Proceedings of 7th international conference on deregulated electricity market issues in South-Eastern Europe (DEMSEE 2012), Nicosia, Sept 2012, pp 1–5

Popock R, Harper J, Ping DC, Jesson J (2015) DECC smart meter small-scale behaviour trials. Synthesis report. Department of Energy and Climate Change, pp 1–93

Prado JSG, Morales WA, Bravo EC, Perez BZ, Reza AE (2020) The power of big data and data analytics for AMI data: a case study. Sensors 20(11):1–27

Price M, Allmeroth T, Cleveland M, Regenwether J (2012) Implementing dynamic pricing—meter configuration trade-offs. Technical report. Deloitte Center for Energy Solutions, pp 1–16

Primadianto A, Lu CN (2017) A review on distribution system state estimation. IEEE Trans Power Syst 32(5):3875–3883

Quiles CG, Exposito AG, Jaén ADLV (2012) State estimation for smart distribution substations. IEEE Trans Smart Grid 3(2):986–995

Rahman M, Mto A (2013) Investigation of bandwidth requirement of smart meter network using OPNET modeler. Smart Grid Renew Energy 4(4):378–390

Ramchurn SD, Vytelingum P, Rogers A, Jennings N (2011) Agent-based control for decentralised demand side management in the smart grid. In: Proceedings of 10th international conference on autonomous agents and multiagent systems (AAMAS), Taipei, May 2011, pp 5–12
Razanousky M, Morrissey K (2018) Fundamental research challenges for distribution state estimation to enable high-performing grids. Final Report, NYSERDA, pp 1–165
Reed M (2018) Understanding how predictive analytics tools benefit power utility asset management. White paper. Schneider Electric, pp 1–5
Rich C, Sisson B, Dasinger A, Chenard M, Atwood G, Eckhart M, Eicher C, Presswood J, Smith P, Hughes M, Gerney A, Buettner S, Ungar L (2013) Residential and commercial buildings. Alliance Commission on National Energy Efficiency Policy, pp 1–44
S3C (2011) Guideline: smart meter monitoring and controlling functionalities. Technical report, pp 1–9
Sachar S, Das S, Emhoff K, Goenka A, Haig K, Pattanaik S, Uchin M (2019) Behavioural energy efficiency potential for India. White paper. Alliance for an Energy Efficient Economy (AEEE), pp 1–35
Sahoo S, Nikovski D, Muso T, Tsuru K (2015) Electricity theft detection using smart meter data. In: Proceedings of IEEE power and energy society innovative smart grid technologies conference (ISGT), Washington, Feb 2015, pp 1–5
Sayed S, Hussain T, Gastli A, Benammar M (2019) Design and realization of an open-source and modular smart meter. Energy Sci Eng 7(4):1405–1422
Schweppe FC, Wildes J (1970) Power system static-state estimation, Part I: exact model. IEEE Trans Power Apparatus Syst PAS-89.1:120–125
Segovia R, Sánchez M (2011) A joint contribution of DG ENER and DG INFSO towards the digital agenda. Action 73: set of common functional requirements of the smart meter. Technical report. European Commission, pp 1–82
Serrenho T, Bertoldi P (2019) Smart home and appliances: state of the art—energy, communications, protocols, standards. JRC technical reports, pp 1–59
Shen B, Ghatikar G, Ni CC, Dudley J, Martin P, Wikler G (2012) Addressing energy demand through demand response: international experiences and practices. Technical report. Environmental Energy Technologies Division, Lawrence Berkeley National Laboratory, pp 1–38
Siano P (2014) Demand response and smart grids—a survey. Renew Sustain Energy Rev 30:461–478
Singh SK, Bose R, Joshi A (2019) Energy theft detection for AMI using principal component analysis based reconstructed data. IET Cyber Phys Syst Theor Appl 4(2):179–185
Souvnik R (2014) Prepaid smart metering. Smart Energy 1(2):42–44
State Government Victoria (2014) Implementing effective energy and water metering systems—the role of metering in managing energy and water consumption. Guidance note. Department of Health, pp 1–13
Thimmapuram PR, Kim J (2013) Consumers' price elasticity of demand modeling with economic effects on electricity markets using an agent-based model. IEEE Trans Smart Grid 4(1):390–397
Toledo F (2013) Smart metering handbook. PennWell Corporation, Tulsa, Oklahoma, pp 1–309
Tounqouet F (2019) European smart metering benchmark—benchmarking smart metering deployment in the EU-28. Tractebel, European Commission DG Energy, pp 1–142
Tram H (2008) Technical and operation considerations in using smart metering for outage management. In: Proceedings of IEEE/PES transmission and distribution conference and exposition, Chicago, April 2008, pp 1–3
Tripathy P (2017) How can utilities benefit from redefining their asset management strategies? Rethinking asset management: evolving to analytics-driven decisions. White paper. Landis+Gyr, pp 1–10
Ulbig A, Zufferey T, Villalon OR, Koch S (2016) Optimized distribution grid operation by utilization of smart metering data. Technical report. Swiss Federal Office of Energy (SFOE), pp 1–59
U.S. DOE (2016) Advanced metering infrastructure and customer systems—results from the smart grid investment grant program. U.S. Department of Energy, Office of Electricity Delivery and Energy Reliability, pp 1–98

USmartConsumer (2014) European smart metering landscape "Utilities and consumers". Technical report. European Union, pp 1–94
Vasak M, Capuder T, Katalin DG, Rácz Á (2018) Deliverable D3.1.1—technology state-of-the art analysis and potential barriers identification for energy management systems in buildings and electricity distribution grids. Project deliverable report. European Union, pp 1–116
Vazquez R, Amaris H, Alonso M, Lopez G, Moreno JI, Olmeda D, Coca J (2017) Assessment of an adaptive load forecasting methodology in a smart grid demonstration project. Energies 10(2):1–23
Vié P, Buvat J, Srivastava A, KVJ S (2015) Big data BlackOut: are utilities powering up their data analytics? Utilities' analytics performance is under-powered. Technical report. Capgemini Consulting, pp 1–14
Vogt G (2020) Energy data innovation network—EDI-Net GA N. 695916, D4.1 overview of smart metering in Germany, Spain and the United Kingdom. Technical report. Energy Data Innovation Network, EDI-Net. Horizon 1–27
Vos LD, Goes M, Melle TV (2018) Consumer satisfaction KPIs for the roll-out of smart metering in the EU member states. Technical report. ASSET project. European Commission, pp 1–113
Waeresch D, Brandalik R, Wellssow WH, Jordan J, Bischler R, Schneider N (2015) State estimation in low voltage grids based on smart meter data and photovoltaic-feed-in-forecast. In: Proceedings of 23rd international conference on electricity distribution (CIRED), Lyon, June 2015, pp 15–18
Waisi ZA, Agyeman MO (2018) On the challenges and opportunities of smart meters in smart homes and smart grids. In: Proceedings of 2nd international symposium on computer science and intelligent control (ISCSIC), Stockholm, Sept 2018, pp 1–6
Wakeel AA, Wu J, Jenkins N (2016) State estimation of medium voltage distribution networks using smart meter measurements. Appl Energy 184:207–218
Wang Y, Chen Q, Hong T, Kang C (2019) Review of smart meter data analytics: applications, methodologies, and challenges. IEEE Trans Smart Grid 10(3):3125–3148
Webborn E, Oreszczyn T (2019) Champion the energy data revolution. Nat Energy 4(8):624–626
Webborn E, Elam S, McKenna E, Oreszczyn T (2019) Utilising smart meter data for research and innovation in the UK. In: Proceedings of ECEEE summer study on energy efficiency, Presqu'ile de Giens, June 2019, pp 1387–1396
Weranga KSK, Kumarawadu S, Chandima DP (2013) Software engineering—from auxiliary to key technologies. In: Springer briefs in applied sciences and technology. Springer, Singapore, pp 1–141
Wijaya TK, Ganu T, Chakraborty D, Aberer K, Seetharam DP (2014) Consumer segmentation and knowledge extraction from smart meter and survey data. In: Proceedings of 2014 SIAM international conference on data mining (SDM), Philadelphia, April 2014, pp 226–234
Wilson C, Hargreaves T, Baldwin RH (2017) Benefits and risks of smart home technologies. Energy Policy 103:72–83
World Bank Group (2018a) Data analytics for advanced metering infrastructure—a guidance note for south Asian power utilities. Technical report. Energy and Extractives Global Practice Group South Asia Region, pp 1–124
World Bank Group (2018b) Survey of International experience in advanced metering infrastructure and its implementation. Technical report, pp 1–104
World Energy Council (2018) The role of ICT in energy efficiency management—household sector. Technical report, pp 1–31
Yang B, Liu S, Gaterell M, Wang Y (2019) Smart metering and systems for low-energy households: challenges, issues and benefits. Adv Build Energy Res 13(1):80–100
Yildiz B, Bilbao JI, Dore J, Sproul AB (2017) Recent advances in the analysis of residential electricity consumption and applications of smart meter data. Appl Energy 208:402–427
Yildiz B, Bilbao JI, Dore J, Sproul AB (2018) Short-term forecasting of individual household electricity loads with investigating impact of data resolution and forecast horizon. Renew Energy Environ Sustain 3(3):1–9

Yuan Y, Dehghanpour K, Bu F, Wang Z (2020) A data-driven customer segmentation strategy based on contribution to system peak demand. IEEE Trans Power Syst 35(5):4026–4035

Zabkowski T, Gajowniczek K (2013) Smart metering and data privacy issues. Inf Syst Manag 2(3):239–249

Zepter JM, Lüth A, Granado PCD, Egging R (2019) Prosumer integration in wholesale electricity markets: synergies of peer-to-peer trade and residential storage. Energy Build 184:163–176

Zhang Y, Huang T, Ettore FB (2018) Big data analytics in smart grids: a review. Energy Inf 1(8):1–24

Zhou L, Xu FY, Ma YN (2010) Impact of smart metering on energy efficiency. In: Proceedings of international conference on machine learning and cybernetics (CMLC), Qingdao, July 2010, pp 3213–3218

Zipperer A, Young PAA, Suryanarayanan S, Roche R, Earle L, Christensen D, Bauleo P, Zimmerle D (2013) Electric energy management in the smart home: perspectives on enabling technologies and consumer behavior. Proc IEEE 101(11):2397–2408

The manufacturer's authorised representative in the EU is Springer Nature Customer Service Centre GmbH, Europaplatz 3, 69115 Heidelberg, Germany. If you have any concerns regarding our products, please contact ProductSafety@springernature.com

Printed and bound by CPI Group (UK) Ltd, Croydon, CR0 4YY
01/10/2026
02225852-0014